A GUIDE
TO HYDROLOGIC
ANALYSIS USING
SCS METHODS

A GUIDE
TO HYDROLOGIC
ANALYSIS USING
SCS METHODS

RICHARD H. McCUEN

University of Maryland

PRENTICE-HALL, INC., Englewood Cliffs, New Jersey 07632

Library of Congress Cataloging in Publication Data

McCuen, Richard H.
    A guide to hydrologic analysis using SCS
methods.

    1. Hydrology—Methodology. I. Title.
GB656.M38   551.48′028   81-8512
ISBN 0-13-370205-7   AACR2

Printed in the United States of America

10 9 8 7 6 5

Prentice-Hall Internation, Inc., *London*
Prentice-Hall of Australia Pty. Limited, *Sydney*
Prentice-Hall of Canada, Ltd., *Toronto*
Prentice-Hall of India Private Limited, *New Delhi*
Prentice-Hall of Japan, Inc., *Tokyo*
Prentice-Hall of Southeast Asia Pte. Ltd., *Singapore*
Whitehall Books Limited, *Wellington, New Zealand*

To

WENDY and ADAM

TABLE OF CONTENTS

LIST OF FIGURES

LIST OF TABLES

LIST OF TABLES (Cont'd)

LIST OF TABLES (Cont'd)

SECTION 1

INTRODUCTION

Recently, there has been a very substantial increase in the number of practicing engineers and hydrologists who are using Soil Conservation Service (SCS) hydrologic methods. As an example, the state of Maryland now requires engineers to use the SCS hydrologic methods on projects requiring state approval. In other states, the SCS methods are widely used both for analyses of small urban watersheds and conservation practices on agricultural watersheds. Additionally, the SCS hydrologic methods are used internationally. These methods are widely used because of their easy-to-apply approach.

The following four SCS publications, which provide the details on hydrologic analyses using the SCS methods, are widely distributed and used:

1. NEH-4: Hydrology, Section 4, National Engineering Handbook

2. TP-149: A Method for Estimating Volume and Rate of Runoff in Small Watersheds

3. TR-20: Computer Program for Project Formulation, Hydrology, Technical Release No.20

4. TR-55: Urban Hydrology for Small Watersheds, Technical Release No. 55

While these SCS publications are readily available, their organization limits their use as either a learning tool or a users manual. Because of their organization, they serve mainly as documentation. This fact, when combined with the significant increase in the use of SCS methods, has created a need for a manual that can be used as a learning aid.

The intent of this guide is to provide both a summary of the basic concepts underlying the SCS hydrologic methods and a guide to their use. Thus, this guide will fill a void that currently exists. It is mainly concerned with the TR-20 method, which is a computer program, and the TR-55 methods, which includes a set of simplified procedures that are usually solved by hand.

Features of the Guide

This guide has numerous features that should make it attractive to both the student or practicing engineer who want to learn the methods and those who need a well organized reference manual. Some of the unique features are:

- computation sheets, which are not contained in the SCS publication, for the manual computation methods of TR-55;

- very basic examples, rather than the extensive examples of the TR-20 publication;

1

- a users manual style that describes the TR-20 input (this is not contained within TR-20);

- discussions within the TR-20 program description that give a direct reference to the methodology and equations that form the basis of the method;

- short sections that isolate information that is specific to each topic, rather than the scattered approach used in NEH-4 and TR-20.

It should be emphasized that this guide does not introduce any new methods, but is designed mainly to present the methodologies in a format that should be more conducive to learning.

Format of the Guide

This guide is separated into four parts, with

Subject	Section(s)
1. General Topics	2-7
2. TR-55: Graph Method	8
Chart Method	9
Tabular Method	10
3. TR-20. Formulation	11-15
Computer Program	16
4. Stormwater Management	17

The general topics of the first part are required for both TR-20 and TR-55 methods, as well as the stormwater detention computations; these topics include runoff volume determination and methods for estimating the time-of-concentration. TR-55 includes descriptions of procedures for estimating the peak discharge; the graph, chart, and tabular methods are discussed in the second part (i.e., sections 8, 9, and 10, respectively). The conceptual framework of TR-20 (sections 11-15) and a user's manual (section 16) represent the third part of this guide. A discussion of the use of TR-55 and TR-20 for sizing stormwater detention basins is provided in part 4 (section 17).

SECTION 2

VOLUME-DURATION-FREQUENCY

Hydrologic analyses must recognize the random nature of precipitation; that is, future occurrences of precipitation can not be predicted. There is a common misconception that if a large flood producing rainfall occurs this year, then a similar rainfall cannot occur next year. To place this concept in the proper perspective, it is reasonable to apply definitions to the appropriate hydrologic concepts.

Most hydrologic analyses involve the concept of volume-duration-frequency. Definitions for these terms are as follows:

Duration: the length of time over which a precipitation event occurs.

Volume: the amount of precipitation occurring over the storm duration.

Frequency: the frequency of occurrence of events having the same volume and duration.

Closely related to these definitions is the concept of intensity, which equals the volume divided by the duration. The volume is often reported as a depth, with units of length such as inches; in such cases, the depth is assumed to occur uniformily over the watershed. Thus, the volume equals the depth times the watershed area.

Just as each concept is important by itself, it is also important to recognize the interdependence of these terms in hydrologic analyses. The volume of precipitation is an upper limit on the amount of runoff. However, a specified volume of runoff may occur from many different combinations of intensities and durations, and these different combinations of intensities and durations will have a significant effect on both runoff volumes and rates. For example, 3 inches of precipitation may result from any of the following combinations of intensity and duration:

Intensity (in/hr)	Duration (hr)	Volume (in)
12	0.25	3
6	0.50	3
3	1.00	3
1.5	2.00	3

Because the rainfall intensity is an important determinant of the hydrologic response, it is important to specify both the volume and duration (or intensity and duration) and not just the total volume.

3

Just as intensity, duration, and volume are interdependent, the fourth concept, frequency, is also a necessary determinant. Frequency can be discussed either in terms of the exceedance probability of the return period, which are defined as follows:

Exceedance probability: the probability that an event having a specified volume and duration will be exceeded in one time period, which is most often assumed to be one year;

Return period: the average length of time between events having the same volume and duration.

The exceedance probability (p) and return period (T) are related by:

$$p = \frac{1}{T} \qquad (1)$$

The following table gives selected combinations of Eq. 1:

T	p
2 yrs	0.50
10	0.10
100	0.01

For example, if a storm of a specified duration and volume has a 1 percent chance of occurring in any one year, then it has an exceedance probability of 0.01 and a return period of 100 years. The argument for not using the return period to interpret the concept of frequency is that it is sometimes improperly interpreted. Specifically, some individuals believe that if a 100-year flood occurs in any one year, then it can not occur for another 100 years; this belief is false because it implies that floods occur deterministically rather than randomly. Because floods occur randomly, there is a finite probability that the 100-year flood could occur in two consecutive years. Thus, the exceedance probability concept is preferred by many.

Events having similar intensities may differ significantly in volume and duration when there is a difference in frequency. For example, the following three storms have similar intensities, but differ significantly in volume, duration, and frequency:

Volume (in)	Duration (hrs)	Frequency (yrs)	Intensity (in/hr)
5.8	8	100	0.72
3.3	4	10	0.82
1.6	2	2	0.80

This example illustrates the need to consider the frequency of the event, as well as the volume, or intensity, and duration.

Because of the importance of the intensity-duration-frequency (IDF) relationship in hydrologic analyses, IDF curves have been compiled for most localities; the IDF curve for Baltimore, Maryland, is shown in Fig. 1.

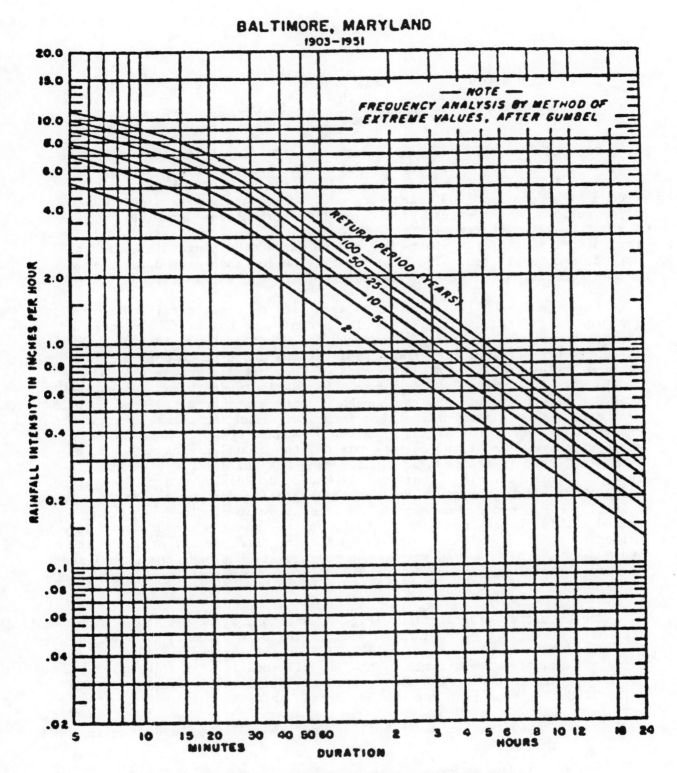

FIGURE 1. RAINFALL INTENSITY-DURATION-FREQUENCY

SOURCE: TECHNICAL PAPER NO. 25
U.S. DEPARTMENT OF COMMERCE
WEATHER BUREAU

5

SECTION 3

THE SCS 24-HOUR STORM DISTRIBUTIONS

The SCS developed two dimensionless rainfall distributions using the
Weather Bureau's Rainfall Frequency Atlases. The rainfall frequency data
for areas less than 400 square miles, for durations to 24 hours, and for
frequencies from 1 to 100 years were used. Data analysis indicated two
major regions, and the resulting rainfall distributions were labeled type
I and type II. The type I distribution is intended for Hawaii, Alaska, and
the coastal side of the Sierra Nevada and Cascade Mountains in California,
Oregon, and Washington. The type II distribution should be used in the remainder of the United States, Puerto Rico, and the Virgin Islands.

The type I and II distributions are based on the generalized rainfall
depth-duration-frequency relationships shown in technical publications of
the Weather Bureau, and rainfall depths for durations from 30-minutes to
24-hours were obtained from these publications and used to derive the two
storm distributions. Using increments of 30 minutes, incremental rainfall
depths were determined. For example, the 30-minute depth was subtracted
from the one-hour depth and the one-hour depth was subtracted from the 1.5-
hour depth. The distributions were formed by arranging these 30-minute incremental depths such that the greatest 30-minute depth is assumed to occur
at about the middle of the 24-hour period, the second largest 30-minute incremental depth in the next 30 minutes, and the third largest in the pre-
ceeding 30 minutes. This continues with each decreasing order of magnitude
until the smaller increments fall at the beginning and end of the 24-hour
rainfall. This procedure results in the maximum 30-minute depth being contained within the maximum 1-hour depth, and the maximum 1-hour depth is contained within the maximum 1.5-hour depth, etc. Because all of the critical
storm depths are contained within the storm distributions, the type I and
II distributions are appropriate for designs on both small and large watersheds.

The resulting distributions (Fig. 2) are most often presented with the
ordinates given on a dimensionless scale. The type I and type II dimensionless distributions plot as a straight line on log-log paper. While they do
not agree exactly with distributions from all locations in the region for
which they are intended, the differences are within the accuracy of the
rainfall depths read from the Weather Bureau atlases.

Fig. 2 shows the two distributions and the ordinates are given in Table
1.

6

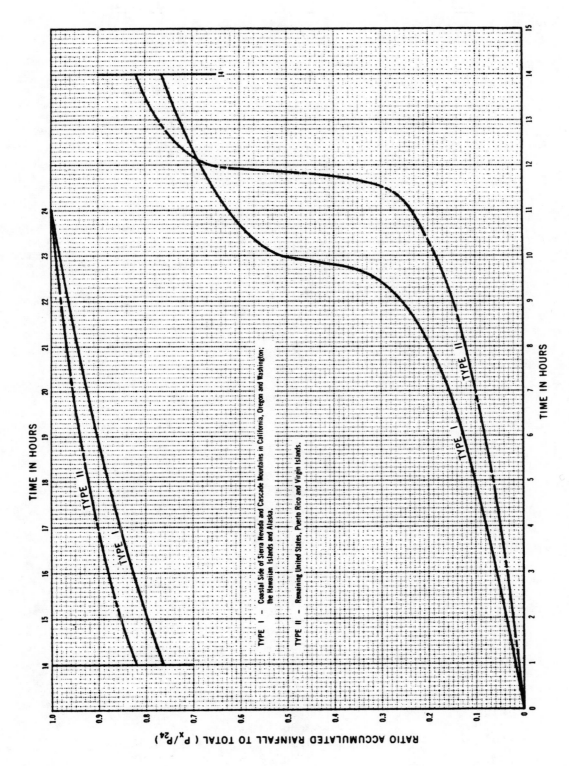

Figure 2. Twenty-four hour rainfall distributions (SCS).

7

TABLE 1. Ordinates of the SCS Type I and Type II Precipitation Distributions

Storm Time (hours)	Precipitation Ratio Type I	Type II
0.0	0.000	0.0000
0.5	0.008	0.0053
1.0	0.017	0.0108
1.5	0.026	0.0164
2.0	0.035	0.0223
2.5	0.045	0.0284
3.0	0.055	0.0347
3.5	0.065	0.0414
4.0	0.076	0.0483
4.5	0.087	0.0555
5.0	0.099	0.0632
5.5	0.122	0.0712
6.0	0.125	0.0797
6.5	0.140	0.0887
7.0	0.156	0.0984
7.5	0.174	0.1089
8.0	0.194	0.1203
8.5	0.219	0.1328
9.0	0.254	0.1467
9.5	0.303	0.1625
10.0	0.515	0.1808
10.5	0.583	0.2042
11.0	0.624	0.2351
11.5	0.654	0.2833
12.0	0.682	0.6632
12.5	0.705	0.7351
13.0	0.727	0.7724
13.5	0.748	0.7989
14.0	0.767	0.8197
14.5	0.784	0.8380
15.0	0.800	0.8538
15.5	0.816	0.8676
16.0	0.830	0.8801
16.5	0.844	0.8914
17.0	0.857	0.9019
17.5	0.870	0.9115
18.0	0.882	0.9206
18.5	0.893	0.9291
19.0	0.905	0.9371
19.5	0.916	0.9446
20.0	0.926	0.9519
20.5	0.936	0.9588
21.0	0.946	0.9653
21.5	0.955	0.9717
22.0	0.965	0.9777
22.5	0.974	0.9836
23.0	0.983	0.9892
23.5	0.992	0.9947
24.0	1.000	1.0000

SECTION 4

THE SCS RAINFALL-RUNOFF RELATION

The volume of runoff (Q) depends on the volume of precipitation (P) and the volume of storage that is available for retention. The actual retention (F) is the difference between the volumes of precipitation and runoff. Furthermore, a certain volume of the precipitation at the beginning of the storm, which is called the initial abstraction (I_a), will not appear as runoff. The SCS assumed the following rainfall-runoff relation, which is shown schematically in Fig. 3:

$$\frac{F}{S} = \frac{Q}{P-I_a} \tag{2}$$

in which S = the potential maximum retention. The actual retention, when the initial abstraction is considered, is:

$$F = (P-I_a) - Q \tag{3}$$

Substituting Eq. 3 into Eq. 2 yields the following:

$$\frac{(P-I_a) - Q}{S} = \frac{Q}{P-I_a} \tag{4}$$

Rearranging Eq. 4 to solve for Q yields:

$$Q = \frac{(P-I_a)^2}{(P-I_a)+S} \tag{5}$$

The factors in Eq. 5 are best understood when placed in the form of a mass curve. Fig. 4 shows a schematic of the mass curve of Q vs. P. The volume of precipitation is separated into the initial abstraction, the retention, and the runoff.

The initial abstraction is a function of land use, treatment, and condition; interception; infiltration; depression storage; and antecedent soil moisture. An empirical analysis was performed for the development of the SCS rainfall-runoff relation, and the following formula was found to be best for estimating I_a:

$$I_a = 0.2 S \tag{6}$$

Research performed since the development of Eq. 6 has suggested that Eq. 6 may not be correct under all circumstances; however, it remains in use until

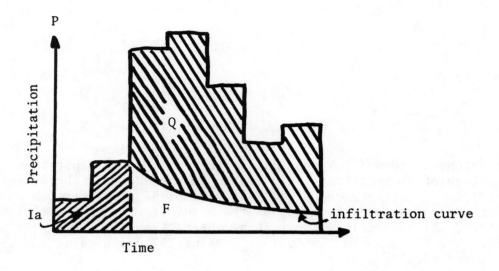

FIGURE 3.   Relationship Between Precipitation, Runoff, and Retention

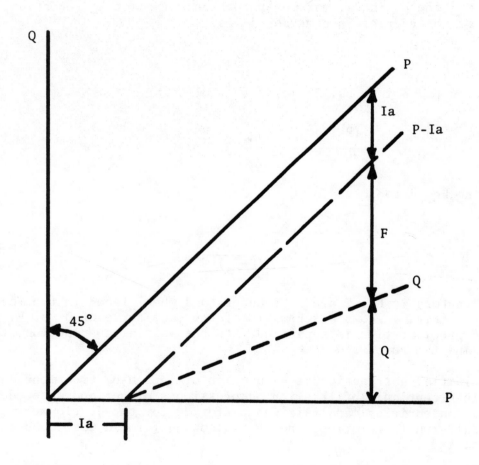

FIGURE 4. A Mass Curve Representation of the SCS Rainfall-Runoff Relationship

a more comprehensive study is accepted. It is important to note that Eq. 6 implies that the factors affecting I_a would also affect S. Substituting Eq. 6 into Eq. 5 yields:

$$Q = \frac{(P-0.2\ S)^2}{P+0.8\ S} \tag{7}$$

While Eq. 5 has two unknowns, I_a and S, Eq. 7 has been reduced to an equation with one unknown, S. Empirical studies indicate that S can be estimated by:

$$S = \frac{1000}{CN} - 10 \tag{8}$$

in which CN = runoff curve number. Thus, the rainfall relationship of Eq. 7, which has one unknown, has been replaced with another relationship with one unknown, CN. Since S is a function of the factors that affect I_a, one should expect that the CN would also be a function of land use, antecedent soil moisture, and other factors that affect runoff and retention.

SECTION 5

CURVE NUMBER ESTIMATION

The volume and rate of runoff depends on both meteorologic and watershed characteristics, and the estimation of runoff requires an index to represent these two factors. The precipitation volume is probably the single most important meteorological characteristic in estimating the volume of runoff. The soil type, land use, and the hydrologic condition of the cover are the watershed factors that will have the most significant impact in estimating the volume of runoff. The antecedent soil moisture will also be an important determinant of runoff volume.

The SCS developed an index, which was called the runoff curve number (CN), to represent the combined hydrologic effect of soil, land use, agricultural land treatment class, hydrologic condition, and antecedent soil moisture. These factors can be assessed from soil surveys, site investigations, and land use maps; when using the SCS hydrologic methods for design the specification of the antecedent soil moisture condition is often a policy decision that suggests average watershed conditions rather than a recognition of a hydrologic condition at a particular time and place.

Soil Group Classification

SCS developed a soil classification system that consists of four groups, which are identified by the letters A, B, C, and D. Soil characteristics that are associated with each group are as follows:

Group A: deep sand, deep loess, aggregated silts

Group B: shallow loess, sandy loam

Group C: clay loams, shallow sandy loam, soils low in organic content, and soils usually high in clay

Group D: soils that swell significantly when wet, heavy plastic clays, and certain saline soils

The SCS soil group can be identified at a site using one of three ways:

1. soil characteristics

2. county soil surveys

3. minimum infiltration rate

The soil characteristics associated with each group are listed above. County soil surveys, where they are made available by Soil Conservation Districts, give a detailed description of the soils at locations within a county; these surveys are usually the best means of identifying the soil group. Soil analyses can be used to estimate the minimum infiltration rates, which can be used to classify the soil using the following values:

Group	Minimum Infiltration Rate (in/hr)
A	0.30 - 0.45
B	0.15 - 0.30
C	0.05 - 0.15
D	0 - 0.05

Cover Complex Classification

The SCS cover complex classification consists of three factors: land use, treatment or practice, and hydrologic condition. There are approximately fifteen different land uses that are identified in the tables for estimating curve number. Agricultural land uses are often subdivided by treatment or practices, such as contoured or straight row; this separation reflects the different hydrologic runoff potential that is associated with variation in land treatment. The hydrologic condition reflects the level of land management; it is separated with three classes: poor, fair, and good. Not all of the land uses are separated by treatment or condition.

Curve Number Estimation

Table 2, which is a compilation of the CN tables provided in NEH-4 and TR-55, show the CN values for the different land uses, treatment, and hydrologic condition; separate values are given for each soil group. For example, the CN for a wooded area with good cover and soil group B is 55; for soil group C, the CN would increase to 70. If the cover (on soil group B) is poor, then the CN will be 66.

Antecedent Soil Moisture Condition

Antecedent soil moisture is known to have a significant effect on both the volume and rate of runoff. Recognizing that it is a significant factor, SCS developed three antecedent soil moisture conditions, which were labeled I, II, and III. The soil condition for each is as follows:

Condition I: soils are dry but not to wilting point; satisfactory cultivation has taken place.

Condition II: average conditions

Condition III: heavy rainfall, or light rainfall and low temperatures have occurred within the last 5 days; saturated soil.

The following table gives seasonal rainfall limits for the three antecedent soil moisture conditions:

	Total 5-day Antecedent Rainfall (inches)	
AMC	Dormant Season	Growing Season
I	Less than 0.5	Less than 1.4
II	0.5 to 1.1	1.4 to 2.1
III	over 1.1	over 2.1

TABLE 2. Runoff Curve Numbers for Hydrologic Soil-Cover Complexes
(Antecedent Moisture Condition II, and $I_a = 0.2\ S$)

Land Use Description/Treatment/Hydrologic Condition			Hydrologic Soil Group			
			A	B	C	D
Residential:[1]						
Average lot size	Average % Impervious[2]					
1/8 acre or less	65		77	85	90	92
1/4 acre	38		61	75	83	87
1/3 acre	30		57	72	81	86
1/2 acre	25		54	70	80	85
1 acre	20		51	68	79	84
Paved parking lots, roofs, driveways, etc.[3]			98	98	98	98
Streets and roads:						
paved with curbs and storm sewers[3]			98	98	98	98
gravel			76	85	89	91
dirt			72	82	87	89
Commercial and business areas (85% impervious)			89	92	94	95
Industrial districts (72% impervious)			81	88	91	93
Open Spaces, lawns, parks, golf courses, cemeteries,etc.						
good condition: grass cover on 75% or more of the area			39	61	74	80
fair condition: grass cover on 50% to 75% of the area			49	69	79	84
Fallow	Straight row	---	77	86	91	94
Row crops	Straight row	Poor	72	81	88	91
	Straight row	Good	67	78	85	89
	Contoured	Poor	70	79	84	88
	Contoured	Good	65	75	82	86
	Contoured & terraced	Poor	66	74	80	82
	Contoured & terraced	Good	62	71	78	81
Small grain	Straight row	Poor	65	76	84	88
		Good	63	75	83	87
	Contoured	Poor	63	74	82	85
		Good	61	73	81	84
	Contoured & terraced	Poor	61	72	79	82
		Good	59	70	78	81
Close -seeded	Straight row	Poor	66	77	85	89
legumes[4]	Straight row	Good	58	72	81	85
or	Contoured	Poor	64	75	83	85
rotation	Contoured	Good	55	69	78	83
meadow	Contoured & terraced	Poor	63	73	80	83
	Contoured & terraced	Good	51	67	76	80
Pasture		Poor	68	79	86	89
or range		Fair	49	69	79	84
		Good	39	61	74	80
	Contoured	Poor	47	67	81	88
	Contoured	Fair	25	59	75	83
	Contoured	Good	6	35	70	79
Meadow		Good	30	58	71	78
Woods or		Poor	45	66	77	83
Forest land		Fair	36	60	73	79
		Good	25	55	70	77
Farmsteads		--- .	59	74	82	86

[1] Curve numbers are computed assuming the runoff from the house and driveway is directed towards the street with a minimum of roof water directed to lawns where additional infiltration could occur.

[2] The remaining pervious areas (lawn) are considered to be in good pasture condition for these curve numbers.

[3] In some warmer climates of the country a curve number of 95 may be used.

[4] Close -drilled or broadcast.

In design, the antecedent soil moisture condition is often a policy decision rather than a statement of actual soil conditions at the site during development.

The CN values obtained from Table 2 are for soil moisture condition II. If either soil condition I or III is to be used, the CN can be adjusted using the following table:

CN for Condition II	Corresponding CN for Condition 1	III
100	100	100
95	87	99
90	78	98
85	70	97
80	63	94
75	57	91
70	51	87
65	45	83
60	40	79
55	35	75
50	31	70
45	27	65
40	23	60
35	19	55
30	15	50
25	12	45
20	9	39
15	7	33
10	4	26
5	2	17
0	0	0

SECTION 6

RUNOFF VOLUME DETERMINATION

Eqs. 7 and 8 provide the means of estimating the volume of runoff when the precipitation volume and the CN are known. This rainfall-runoff relation can also be presented using the mass curve concept presented in Fig. 5. That is, the graphical solution of Fig. 5 can be used in place of the mathematical formulas.

Example 6-1

Determine the runoff volume for the 24-hour, 100-year precipitation of 7 inches when the soil is of group B and the watershed is in row crop, contoured in good hydrologic condition.

The CN from Table 2 yields a CN of 75. Using either the mathematical formulas or the mass curve of Fig. 5 provides a runoff volume of about 4.15 inches.

Example 6-2

Determine the runoff volume for the 24-hour, 100-year precipitation for antecedent soil moisture condition II, with the following land uses and soil groups:

Area Fraction	Land Use/Condition	Soil Group
0.40	meadow: good condition	D
0.25	wooded: poor cover	C
0.20	open space: good condition	D
0.15	residential (1/4 acre lots)	C

Table 2 provides the following CN's for the four land uses: 78, 77, 80, 83, respectively. Therefore, a weighted CN is computed:

$$CN = 0.40(78) + 0.25(77) + 0.20(80) + 0.15(83)$$
$$= 78.9 \text{ (use 79)}$$

For a CN of 79 and a precipitation of 7 inches, the runoff volume is 4.58 inches.

Example 6-3

Fig. 6 shows a watershed that is 40 percent wooded (good condition) and 60 percent residential (1/4 acre lots). The watershed has 75 percent soil group B and 25 percent soil group C. Determine the runoff volume if the precipitation is 7 inches.

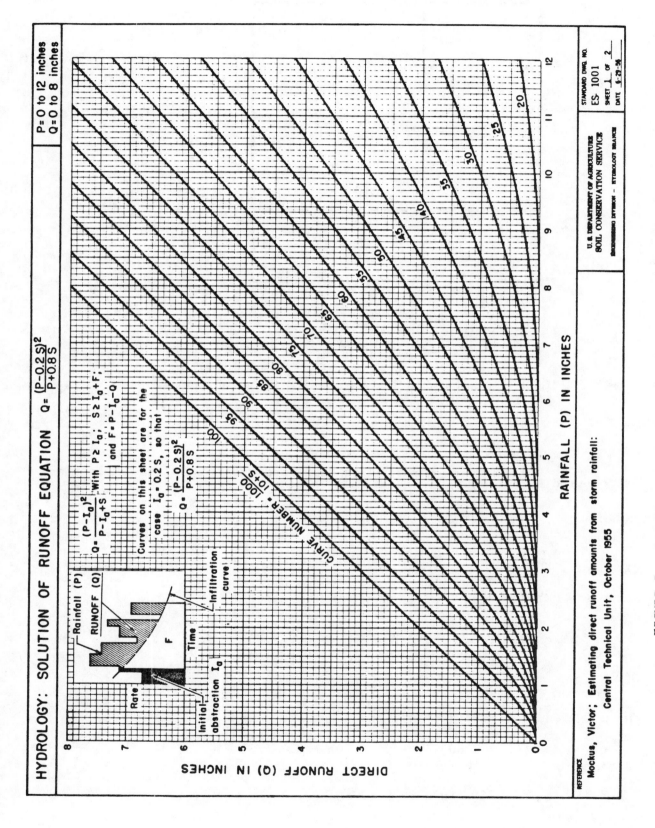

FIGURE 5. Graphical Solution of Rainfall-Runoff Equation

17

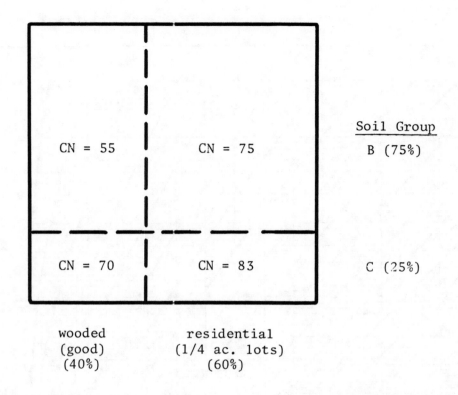

CN = 55   CN = 75   B (75%)

CN = 70   CN = 83   C (25%)

wooded   residential
(good)   (1/4 ac. lots)
(40%)   (60%)

FIGURE 6. Watershed for Estimating Weighted Curve Number

The fraction of the watershed in each land use/soil group combination is computed as follows:

Land Use	Soil Group	Area	CN
wooded	B	0.4(0.75) = 0.30	55
	C	0.4(0.25) = 0.10	70
residential	B	0.6(0.75) = 0.45	75
	C	0.6(0.25) = 0.15	83

The weighted CN is:

$$CN = 0.30(55) + 0.10(70) + 0.45(75) + 0.15(83)$$
$$= 69.7 \text{ (use 70)}$$

The resulting runoff volume is 3.62 inches.

SECTION 7

ESTIMATING THE TIME-OF-CONCENTRATION

Time is an important element in hydrologic forecasting. This is re-flected in the fact that most hydrologic methods include a time variable as input. The SCS methods are no different, and the time-of-concentration was selected as the best indicator of the effects of time.

The time-of-concentration (t_c) is a measure of the time for a particle of water to travel from the hydrologically most distant point in the water-shed to the point where the design is to be made. Additionally, the follow-ing operational definition is sometimes used with respect to unit hydro-graphs: the time-of-concentration is the time from the end of rainfall ex-cess to the point of inflection on the recession. While this operational definition will be used in developing the SCS unit hydrograph, the former definition should be understood for the computation of time-of-concentra-tion estimates.

Hydrologists have developed numerous methods for estimating the time-of-concentration. Two methods are recommended within NEH-4 and TR-55, the lag method and the upland, or velocity, method. Almost all methods of esti-mating the time-of-concentration use the slope, the hydraulic length, and some measure of land use; the lag and velocity methods are no different in that they use these three factors. The hydraulic length is the distance from the hydrologically most distant point in the watershed to the point where the design is to be made.

The Lag Method

The lag method relates the time lag (L) which is defined as the time in hours from the center of mass of rainfall excess to the peak discharge, to the slope (Y) in percent, the hydraulic length (ℓ) in feet, and the maximum retention (S):

$$L = \frac{\ell^{0.8}(S+1)^{0.7}}{1900\,Y^{0.5}} \tag{9}$$

in which S is given by Eq. 8. The time lag can also be determined from the nomograph of Fig. 7. Empirical evidence used in developing the SCS methods resulted in the following relationship between the time-of-concentration and the lag:

$$t_c = \frac{5}{3} L \tag{10}$$

in which t_c is measured in hours.

19

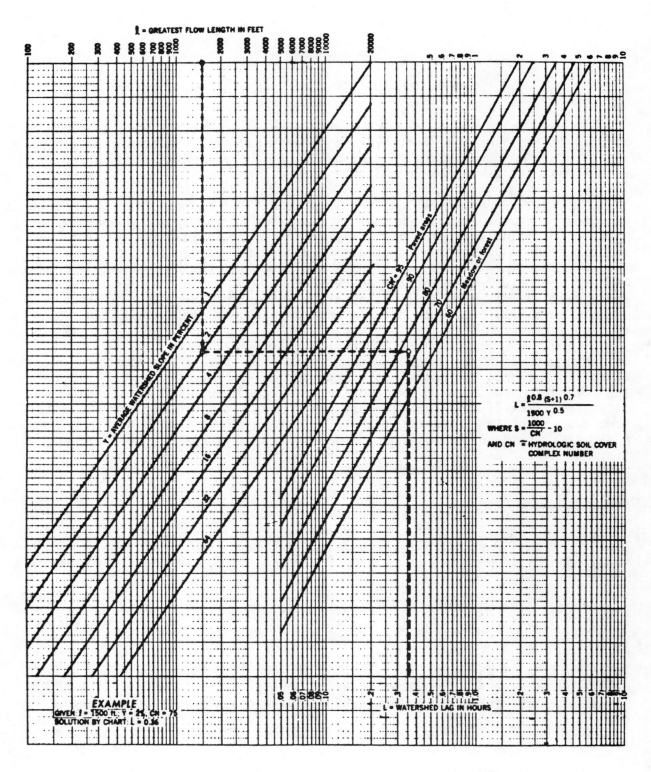

Figure 7.--Curve number method for estimating lag (L)

20

The Upland Method

Essentially the same input is required for estimating t_c with the upland, or velocity, method as with the lag method. The velocity method has an intermediate step in which the velocity is estimated with the land use and the slope; Fig. 8 is used to estimate the velocity in feet per second (fps). The time-of-concentration equals the ratio of the hydraulic flow length to the velocity:

$$t_c = \frac{\ell}{V} \tag{11}$$

If ℓ is measured in feet and V in fps, then the value resulting from Eq. 11 must be divided by 3600 in order to convert t_c from seconds to hours.

Adjustment for Urbanization

The curve number appears not to adequately reflect the effect of the soil cover complex in urban areas on the runoff potential. For composite land use areas where urban land uses provide a more efficient flow pattern than pervious land uses, Eq. 9 overestimates lag. TR-55 provides two figures that can be used to adjust the lag computed by Eq. 9 for the percentages of the hydraulic length that is modified (PHLM) and the impervious areas (PIMP); the lag adjustment factors are used independently, and therefore, both values can be applied on the same project. The lag factor (LF) for each adjustment can be computed using values from the following Equation:

$$LF = 1 - PRCT(-0.006789 + 0.000335\,CN - 0.0000004298\,CN^2 - 0.00000002185\,CN^3)$$

in which CN is the curve number for future land use conditions and PRCT is either the percent hydraulic length modified (PHLM) or the percent impervious area (PIMP). The lag computed from Eq. 9 is then multiplied by the lag factor from the above equation. If both adjustments are necessary, then the equations are used for both modifications and two lag factors are applied to the lag computed from Eq. 9.

TR-55 provides two important guidelines for use of these adjustment factors:

1. Since the lag factors are used only with future-condition curve numbers, the lag factors cannot be used to directly compute the decrease in lag from present conditions.

2. When only peak discharges are to be computed using the TR-55 methods, lag does not have to be computed; therefore, these lag factor adjustments are not necessary.

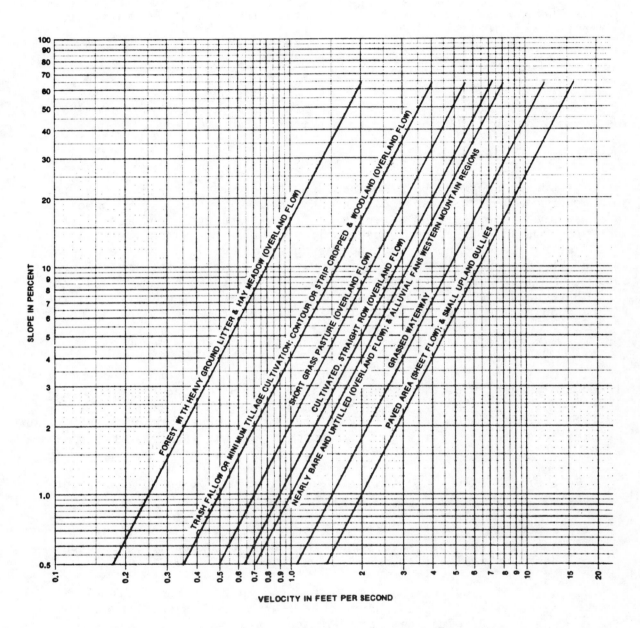

Figure 8. Velocities for upland method of estimating T_c

22

SECTION 8

THE GRAPHICAL METHOD

Part of Chapter 5 in TR-55 describes a method for estimating the peak discharge. The method, which is referred to as the graphical method, derives its name from a graph that relates the time-of-concentration (hours) and the unit peak discharge ($cfs/mi^2/in$). The input data requirements are minimal and include the return period in years (T), the 24-hour, T-year precipitation in inches (P), the runoff curve number (CN), the drainage area in square miles (A), the slope in percent (Y), and the hydraulic length in feet (HL or ℓ). The land use is also required if the velocity method is used to estimate the time-of-concentration.

The procedure requires the volume of runoff to be estimated from either Eq. 7 or Fig. 5 using T, P, and CN as input. The time-of-concentration can be estimated using either the lag method or the velocity method. The unit peak discharge is estimated from Fig. 9. The peak discharge equals the product of the unit peak discharge, the drainage area, and the volume of runoff.

The computation sheet of Table 3 provides a convenient means of summarizing the input data and the resulting peak discharge.

The graphical method is recommended: 1) where valley routing is not required, and 2) for watersheds where land use, soil, and cover are uniformly distributed throughout the watershed.

Example 8-1

Find the peak discharge for a 300 acre watershed having a slope of 4 percent, a CN of 70, and a hydraulic length of 6400 feet. Assume the 24-hour, 25-year precipitation is 5 inches.

The computations are summarized on the computation sheet of Table 4. The lag method yielded a lag of 0.95 hours and a time-of-concentration of 1.58 hours. The runoff volume for a precipitation of 5 inches and a CN of 70 is 2.05 inches. The unit peak discharge for a time-of-concentration of 1.58 hours is 228 $cfs/mi^2/in$. The resulting peak discharge is 219 cfs.

Accuracy of the Graphical Method

The graphical method was developed from the tabular method of Section 10 for the case where the travel time was zero. Thus, the graphical method is subject to the limitations of the tabular method. Specifically, the graphical method should not be used when runoff volumes are less than about 1.5 inches for curve numbers less than 60. Furthermore, the subareas should be less than 20-square miles in area and there should not be large changes in runoff curve numbers among the subareas of the watershed. These limitations are given in TR-55.

TABLE 3. COMPUTATION SHEET: TR-55 GRAPH METHOD

1. Estimate the volume of runoff

   *a. T = _____ (years): return period for design

   *b. P = _____ (inches): 24-hr, T-year precipitation volume (i.e., depth)

   *c. CN = _____ : runoff curve number

   d. Q = _____ (inches): runoff volume obtained from Eq. 7 or Fig. 5

2. Drainage Area: A = _____ (Square miles)

3. Estimate Time-of-Concentration (use either the lag method or the velocity method)

   LAG METHOD

   *a. CN = _____

   *b. Slope = _____ (%)

   *c. hydraulic length = _____ (ft)

   d. L = _____ (hours): from Fig. 7

   e. t_c = _____ (hours) $= \frac{5}{3} L$

   VELOCITY METHOD

   *a. land use _____

   *b. slope = _____ (%)

   *c. hydraulic length (HL) = _____ (ft)

   d. velocity (V) = _____ (fps): from Fig. 8

   e. $t_c = \frac{HL}{3600V}$ _____ (hours)

4. Estimate unit peak discharge (q'_p) = _____ (cfs/mi^2/in): use Fig. 9

5. Estimate peak discharge $q_p = q'_p$ AQ = _____ (cfs)

_____
* indicates required input

24

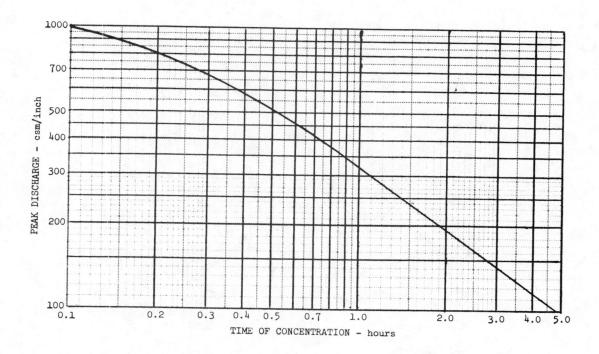

Figure 9. Peak discharge in csm per inch of runoff versus time of concentration (T_c) for 24-hour, type-II storm distribution.

To further define limitations on the graphical method the results of numerous TR-20 runs were compared with estimates of peak discharge made with the graphical method. The runs were made for ranges of the time of concentration (hours), the precipitation volume (inches), and the curve number of 0.5 to 5.0 hours, 1.0 to 10.0 inches, and 50 to 95 curve number units, respectively. The results indicate that the graphical method is a valid approximation of TR-20 as long as the initial abstraction is less than 25 percent of the total 24-hour rainfall; this constraint is easily assessed using the following tabular representation of the constraint, which relates the curve number (CN) and the minimum precipitation:

CN	minimum precipitation
50	8.00 inches
60	5.33
70	3.42
80	2.00
90	0.88
95	0.42

TABLE 4. Example

1. Estimate the volume of runoff

   *a. T = __25__ (years): return period for design

   *b. P = __5__ (inches): 24-hr, T-year precipitation volume (i.e., depth)

   *c. CN = __70__ : runoff curve number

   d. Q = __2.04__ (inches): runoff volume obtained from Eq. 7 or Fig. 5

2. Drainage Area: A = __0.4688__ (Square miles)

3. Estimate Time-of-Concentration (use either the lag method or the velocity method)

   LAG METHOD

   *a. CN = __70__

   *b. Slope = __4__ (%)

   *c. hydraulic length = __6400.__ (ft)

   d. L = __0.95__ (hours): from Fig. 7

   e. t_c = __1.58__ (hours) = $\frac{5}{3}$ L

   VELOCITY METHOD

   *a. land use _____

   *b. slope = _____ (%)

   *c. hydraulic length (HL) = _____ (ft)

   d. velocity (V) = _____ (fps): from Fig. 8

   e. $t_c = \dfrac{HL}{3600V}$ _____ (hours)

4. Estimate unit peak discharge (q'_p) = __228__ (cfs/mi^2/in): use Fig. 9

5. Estimate peak discharge $q_p = q'_p$ AQ = __219__ (cfs)

_____ indicates required input

For precipitation volumes greater than the value given in this table the graphical method should provide reasonable estimates when compared with TR-20. For precipitation volumes less than the stated amount the initial abstraction will be greater than 25 percent of the total rainfall and the method will provide biased estimates of peak discharge.

SECTION 9

THE TR-55 CHART METHOD

Another procedure for computing the peak discharge, which is called the chart method, is described in TR-55. The chart method was designed for use in estimating the effect of development on the peak discharge rate. The input data includes: the hydraulic length (ft), drainage area (acres), the percentage of ponds and swampy area, the design return period (years), the CN, the watershed slope (%), the precipitation depth (inches) for the design return period T, and the percentages of both the impervious area and the hydraulic length modified. All of the data are not necessary for all cases because some of the options are not mandatory. Application of the method is limited to watersheds from 1 acre to 2000 acres. The method is based on a 24-hour storm volume and a type II storm distribution.

An estimate of the peak discharge rate is easily obtained using the computation sheet of Table 5.

The hydraulic length (HL) is used when it is desired to make a shape adjustment. The hydraulic length is entered in Fig. 10 and the effective area (EA) is obtained and entered on the computation sheet. Alternatively, the effective area can be computed using the equation:

$$EA = 0.00013586 \ HL^{5/3} \tag{12}$$

If a watershed shape adjustment is not desired, then the HL is not necessary and EA should be set equal to the drainage area A.

If a significant portion of the watershed is swampy and/or in ponds, the pond and swamp adjustment factor PF can be obtained from Table 6. The value depends on the location of the ponds or swampy area within the watershed, the return period T (or storm frequency), and the percentage of ponding and swampy area PPS.

The unit discharge, which will be discussed below, is obtained from charts designed for index slopes of 1%, 4%, and 16%. For slopes other than these three index values, the slope adjustment factor SF can be obtained from Table 7. The following table indicates the slope designations:

Slope Designation	Index Slope	Slope Range
flat	1%	SP $\leq$ 2.5%
moderate	4%	2.5< SP $\leq$ 7.5%
steep	16%	7.5< SP

The effective area EA and slope SP are used as input to the appropriate part of Table 7, and the slope adjustment factor SF is obtained and recorded on the computation sheet.

TABLE 5.   COMPUTATION SHEET FOR CHART METHOD

PROJECT _____   Computed By _____ Date _____
       _____   Checked By _____ Date _____

1. Required Input

    A = _____ Acres : Drainage Area
    T = _____ Years : Design Frequency (return period)
    P = _____ Inches: Rainfall depth for 24-hour, T-year event
    Y = _____ %   : Average watershed slope
    CN = _____   : Runoff Curve Number

2. Compute Volume of Runoff, Q

    Q = _____ Inches: Use CN and P as input to Fig. 5

3. Watershed Shape Adjustment (Optional: if adjustment is not made, set
   EA = A)

    HL = _____ feet : Hydraulic Length
    EA = _____ Acres : Equivalent Drainage Area (use Fig. 10)
    HF = _____   : HF = A/EA

4. Obtain Unit Peak Discharge, QU

    QU = _____ cfs/inch Q : Use EA with Fig. 11 (Sheet 1, 2, and 3 for
                                 flat, moderate, and steep slopes, respectively)

5. Watershed Slope Interpolation Factor, SF (Optional: if adjustment is not
   made, set SF = 1.0)

    SF = _____   : Use Y and EA with Table 7

6. Ponding and Swamp Storage Adjustment Factor, PF (Optional: if adjustment
   is not made, set PF = 1.0)

    PPS = _____ % : % of Ponds and Swampy Area (Based on actual drainage
                             area A)
    Location in watershed (check one):
     Design Point (6-a)___; Center or Spread out (6-b)___; Upper Reaches (6-c)___
    PF = _____   : Use PPS and T with Table 6-a, 6-b, or 6-c.

7. Peak Discharge QP, Calculation with Adjustments
    QP =   QU   x   Q   x   HF   x   SF   x   PF
       = _____ x _____ x _____ x _____ x _____
       = _____cfs

8. Modifications for Urbanization
    IMP = _____ % : Percentage of Impervious Area (based on actual
                             drainage area A)
    IMPF = _____   : Impervious Area Adjustment Factor (Fig. 12)
    HLM = _____ % : Percentage of Hydraulic Length Modified
    HLMF = _____   : Hydraulic Length Modified Factor (Fig. 13)
    QPU =   QP   x   IMPF   x   HLMF
        = _____ x _____ x _____
        = _____cfs

29

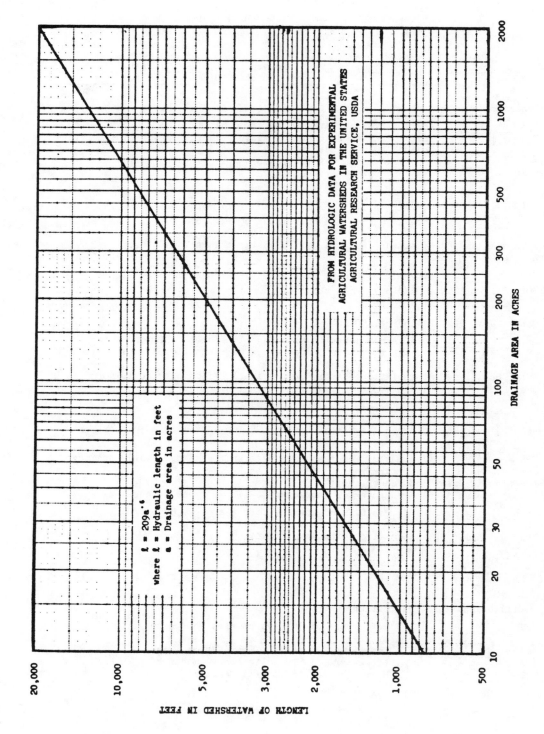

Figure 10.--Hydraulic length and drainage area relationship.

30

TABLE 6. **Adjustment factors where ponding and swampy areas occur at the design point**

Ratio of drainage area to ponding and swampy area	Percentage of ponding and swampy area	Storm frequency (years)					
		2	5	10	25	50	100
500	0.2	0.92	0.94	0.95	0.96	0.97	0.98
200	.5	.86	.87	.88	.90	.92	.93
100	1.0	.80	.81	.83	.85	.87	.89
50	2.0	.74	.75	.76	.79	.82	.86
40	2.5	.69	.70	.72	.75	.78	.82
30	3.3	.64	.65	.67	.71	.75	.78
20	5.0	.59	.61	.63	.67	.71	.75
15	6.7	.57	.58	.60	.64	.67	.71
10	10.0	.53	.54	.56	.60	.63	.68
5	20.0	.48	.49	.51	.55	.59	.64

Adjustment factors where ponding and swampy areas are spread throughout the watershed or occur in central parts of the watershed

Ratio of drainage area to ponding and swampy area	Percentage of ponding and swampy area	Storm frequency (years)					
		2	5	10	25	50	100
500	0.2	0.94	0.95	0.96	0.97	0.98	0.99
200	.5	.88	.89	.90	.91	.92	.94
100	1.0	.83	.84	.86	.87	.88	.90
50	2.0	.78	.79	.81	.83	.85	.87
40	2.5	.73	.74	.76	.78	.81	.84
30	3.3	.69	.70	.71	.74	.77	.81
20	5.0	.65	.66	.68	.72	.75	.78
15	6.7	.62	.63	.65	.69	.72	.75
10	10.0	.58	.59	.61	.65	.68	.71
5	20.0	.53	.54	.56	.60	.63	.68
4	25.0	.50	.51	.53	.57	.61	.66

Adjustment factors where ponding and swampy areas are located only in upper reaches of the watershed

Ratio of drainage area to ponding and swampy area	Percentage of ponding and swampy area	Storm frequency (years)					
		2	5	10	25	50	100
500	0.2	0.96	0.97	0.98	0.98	0.99	0.99
200	.5	.93	.94	.94	.95	.96	.97
100	1.0	.90	.91	.92	.93	.94	.95
50	2.0	.87	.88	.88	.90	.91	.93
40	2.5	.85	.85	.86	.88	.89	.91
30	3.3	.82	.83	.84	.86	.88	.89
20	5.0	.80	.81	.82	.84	.86	.88
15	6.7	.78	.79	.80	.82	.84	.86
10	10.0	.77	.77	.78	.80	.82	.84
5	20.0	.74	.75	.76	.78	.80	.82

Table 7 .--Slope adjustment factors by drainage areas

FLAT SLOPES

Slope (per-cent)	10 acres	20 acres	50 acres	100 acres	200 acres	500 acres	1,000 acres	2,000 acres
0.1	0.49	0.47	0.44	0.43	0.42	0.41	0.41	0.40
0.2	.61	.59	.56	.55	.54	.53	.53	.52
0.3	.69	.67	.65	.64	.63	.62	.62	.61
0.4	.76	.74	.72	.71	.70	.69	.69	.69
0.5	.82	.80	.78	.77	.77	.76	.76	.76
0.7	.90	.89	.88	.87	.87	.87	.87	.87
1.0	1.00	1.00	1.00	1.00	1.00	1.00	1.00	1.00
1.5	1.13	1.14	1.14	1.15	1.16	1.17	1.17	1.17
2.0	1.21	1.24	1.26	1.28	1.29	1.30	1.31	1.31

MODERATE SLOPES

Slope (per-cent)	10 acres	20 acres	50 acres	100 acres	200 acres	500 acres	1,000 acres	2,000 acres
3	.93	.92	.91	.90	.90	.90	.89	.89
4	1.00	1.00	1.00	1.00	1.00	1.00	1.00	1.00
5	1.04	1.05	1.07	1.08	1.08	1.08	1.09	1.09
6	1.07	1.10	1.12	1.14	1.15	1.16	1.17	1.17
7	1.09	1.13	1.18	1.21	1.22	1.23	1.23	1.24

STEEP SLOPES

Slope (per-cent)	10 acres	20 acres	50 acres	100 acres	200 acres	500 acres	1,000 acres	2,000 acres
8	.92	.88	.84	.81	.80	.78	.78	.77
9	.94	.90	.86	.84	.83	.82	.81	.81
10	.96	.92	.88	.87	.86	.85	.84	.84
11	.96	.94	.91	.90	.89	.88	.87	.87
12	.97	.95	.93	.92	.91	.90	.90	.90
13	.97	.97	.95	.94	.94	.93	.93	.92
14	.98	.98	.97	.96	.96	.96	.95	.95
15	.99	.99	.99	.98	.98	.98	.98	.98
16	1.00	1.00	1.00	1.00	1.00	1.00	1.00	1.00
20	1.03	1.04	1.05	1.06	1.07	1.08	1.09	1.10
25	1.06	1.08	1.12	1.14	1.15	1.16	1.17	1.19
30	1.09	1.11	1.14	1.17	1.20	1.22	1.23	1.24
40	1.12	1.16	1.20	1.24	1.29	1.31	1.33	1.35
50	1.17	1.21	1.25	1.29	1.34	1.37	1.40	1.43

The unit discharge is then obtained from Fig. 11, which is separated on the basis of the three index slopes. The unit discharges are given with units of cfs/inch of runoff. Fig. 11 defines the unit discharges for the SCS type II storm. The CN is used with the effective area EA to get the unit discharge.

Using the depth of precipitation and the CN, the volume of runoff (inches) can be determined from Fig. 5.

If an adjustment is to be made for the percentage of imperviousness, the peak adjustment factor FIMP is obtained from Fig. 12. The percentage of imperviousness IMP and the CN are used as input.

A similar adjustment is used when the hydraulic flow pattern has been or will be modified. The percentage of the hydraulic length modified HLM and the CN are used as input to Fig. 13 to get the adjustment factor HLMF.

Example 9-1

Estimate the effect on the 50-year peak discharge of modifying 30 percent of the hydraulic flow length of a 300 acre watershed. The watershed has an average slope of 4 percent and a CN of 70. Assume that there is no impervious area or ponds and swampy area. The following table shows the computations for the before-development condition. Also assume that the 50-year precipitation depth is 5 inches.

1.	A = 300 acres	2.	RO = 2.05 inches	5.	SF = 1.0	
	T = 50 years					
	P = 5 inches	3.	HL = 0 feet	6.	PPS = 0	
	Y = 4%		EA = 300 acres		PF = 1.0	
	CN = 70		HF = 1.0			
				7.	QP = 225 cfs	
		4.	QU = 110 cfs/in			

If 30 percent of the hydraulic length is modified, then the adjustment factor HLFM from Fig. 13 equals 1.22. Thus, the peak will increase by 22 percent to 275 cfs.

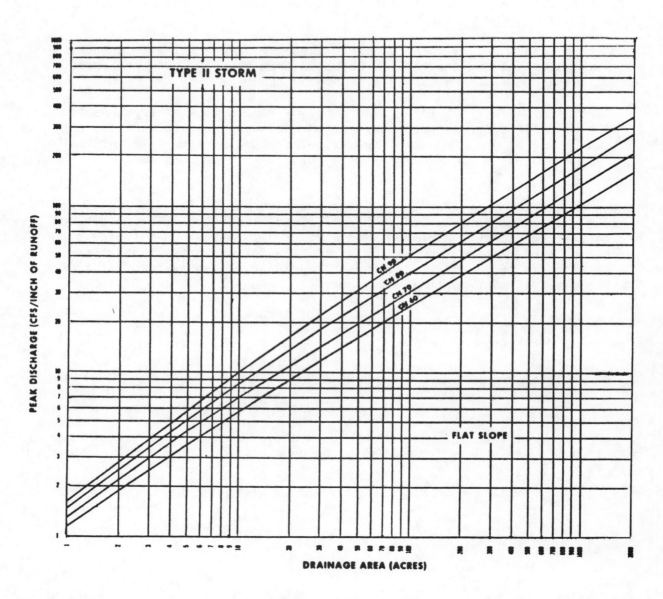

Figure 11.—Peak rates of discharge for small watersheds (24-hour, type-II storm distribution). Sheet 1 of 3

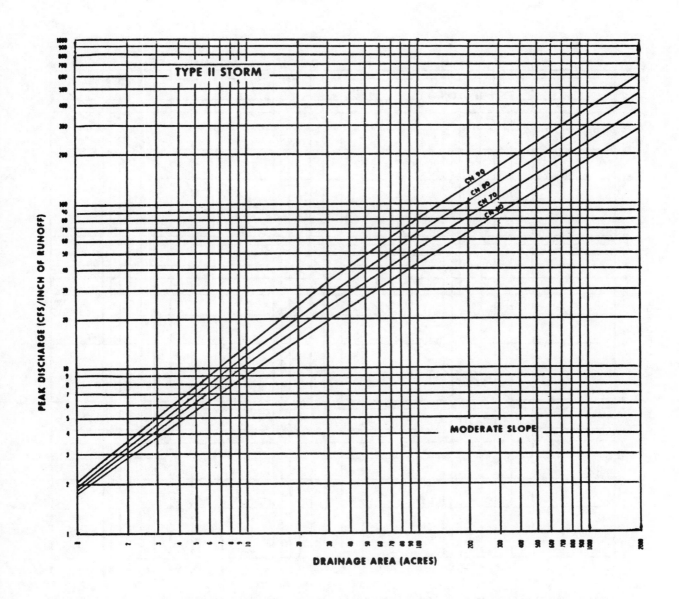

Figure 11.—Peak rates of discharge for small watersheds (24-hour, type-II storm distribution). Sheet 2 of 3

35

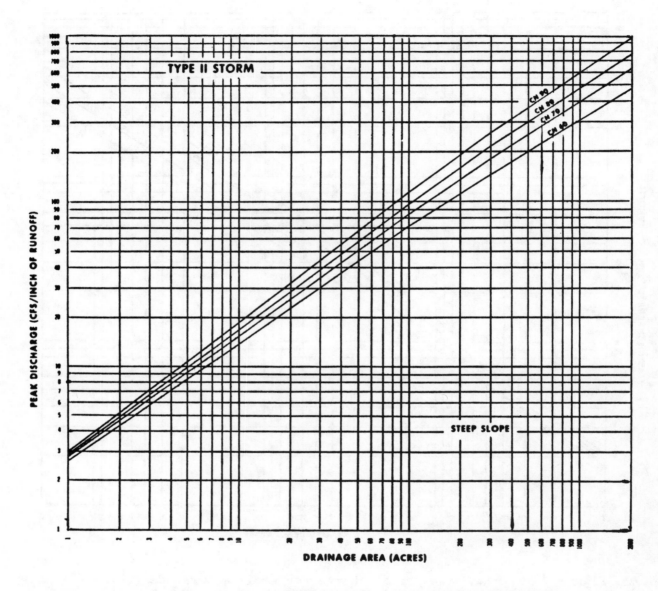

Figure 11.--Peak rates of discharge for small watersheds (24-hour, type-II storm distribution). Sheet 3 of 3

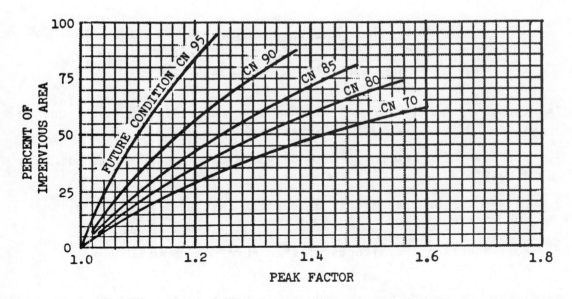

Figure 12.—Factors for adjusting peak discharges for a given future-condition runoff curve number based on the percentage of impervious area in the watershed.

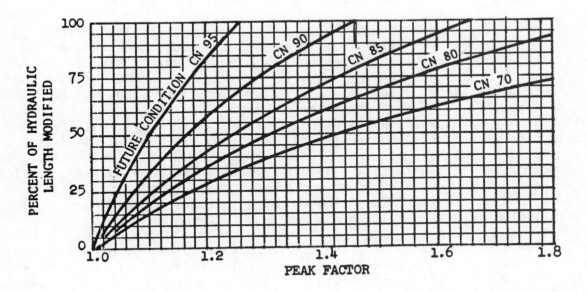

Figure 13.—Factors for adjusting peak discharges for a given future-condition runoff curve number based on the percentage of hydraulic length modified.

SECTION 10

THE TABULAR METHOD OF TR-55

The tabular method, which is discussed in Chapter 5 of TR-55, was designed for use in the following circumstances:

1. For developing composite flood hydrographs at any point within a watershed;

2. For measuring the effects of changes in land use of one or more subwatersheds; and

3. For assessing the effects of structures or combinations of structures.

In general, the procedure was intended for measuring the effect on the composite flood hydrograph of changes within subwatersheds of a larger drainage area.

The input requirements for the tabular method are minimal. The 24-hour rainfall depth (inches) for a selected exceedence probability (i.e., return period) is required. For each subwatershed, the drainage area, the runoff curve number, and the time-of-concentration must be determined. Additionally, the travel time for each channel reach is necessary.

Before the method is used, a user should be familiar with several constraints. First, the constraints that were used in developing the method are important when applying the method. The tabular method was developed by making numerous computer runs with the TR-20 program. In each case, a runoff CN of 75 was used and the rainfall volumes were sufficient to yield 3 inches of runoff. When the tabular method is applied to cases having characteristics that are significantly different from the conditions used in developing the method, then the resulting hydrograph may not provide close agreement with the hydrograph that would result from a TR-20 analysis. These assumptions are not considered to be critical when the sole purpose in using the method is to assess the effect of changes in a watershed, such as land use or structure changes. The difference in the before/after hydrographs is relatively insensitive to the assumption of a CN of 75.

In order to make accurate assessments of watershed changes, there are certain limitations that should be adhered to in applying the method. First, within any subwatershed there should be little variation in CN; this does not mean that subwatersheds should have similar CN's but that each subwatershed should have little variation in soil and land use characteristics. Second, the area of each subwatershed should be less than 20 square miles. Third, the precipitation should be sufficient to yield runoff volumes greater than 1.5 inches, especially when CN's are less than 60.

The solution methodology centers about the tabular discharge values of

Table 8.—Tabular discharges for type-II storm distribution (csm/in)

TIME OF CONCENTRATION = 0.1 hours
Hydrograph Time in Hours

t_t	11.0	11.5	11.7	11.8	11.9	12.0	12.1	12.2	12.3	12.4	12.5	12.6	12.7	12.8	12.9	13.0	13.2	13.5	14.0	14.5	15.0	16.0	18.0	20.0
0	24	51	299	991	746	477	233	152	132	121	111	85	74	70	68	65	52	48	39	33	29	24	18	14
0.25	20	38	66	140	327	626	686	546	354	236	169	137	117	97	83	75	66	52	41	35	30	24	18	14
0.50	15	27	36	43	67	133	288	482	580	543	429	310	222	168	134	110	81	63	47	32	32	26	19	15
0.75	12	20	25	29	34	42	65	125	245	392	496	515	452	360	273	206	127	80	53	42	35	27	19	15
1.00	9	15	19	21	28	20	32	41	63	115	209	328	427	470	451	389	245	121	64	38	38	29	20	16
1.50	6	10	12	13	14	16	17	19	22	25	29	38	56	92	154	236	410	360	133	66	47	33	21	17
2.00	3	6	7	8	9	10	11	12	13	14	16	18	20	23	27	34	74	244	371	142	68	38	23	19
2.50	2	4	4	5	6	6	7	7	8	9	10	11	12	13	15	16	21	41	243	343	150	48	26	20
3.00	1	2	2	3	4	4	4	4	5	5	6	7	7	8	9	10	12	17	50	239	321	74	29	21
3.50	0	1	1	1	2	2	2	2	3	3	4	4	4	5	6	6	7	10	17	59	304	159	33	23
4.00	0	0	0	0	1	1	1	1	1	2	2	2	2	3	3	4	5	6	10	18	67	290	39	23

TIME OF CONCENTRATION = 0.2 hours
Hydrograph Time in Hours

t_t	11.0	11.5	11.7	11.8	11.9	12.0	12.1	12.2	12.3	12.4	12.5	12.6	12.7	12.8	12.9	13.0	13.2	13.5	14.0	14.5	15.0	16.0	18.0	20.0
0	23	47	208	509	796	641	424	245	170	138	121	104	85	75	71	68	56	49	40	34	29	24	18	16
0.25	18	34	49	91	196	419	603	627	486	341	235	173	138	114	96	83	70	55	43	36	31	25	18	15
0.50	14	24	32	37	50	67	161	341	490	545	497	397	296	219	167	133	92	67	49	33	33	26	19	15
0.75	11	18	23	26	30	36	49	84	161	284	409	491	481	422	340	263	157	89	56	36	36	27	19	15
1.00	9	14	18	20	22	25	29	35	48	79	143	240	347	426	452	427	299	147	69	39	39	29	20	16
1.50	5	9	11	12	13	14	16	18	20	23	26	32	43	67	110	176	330	399	159	72	50	33	22	17
2.00	3	6	7	7	8	9	10	11	12	13	15	16	18	21	24	29	56	192	363	168	75	40	24	18
2.50	1	3	4	5	6	6	6	7	7	8	9	10	11	12	13	15	19	33	200	337	174	51	26	19
3.00	0	2	2	2	3	3	4	4	5	5	6	6	7	8	9	9	11	15	40	203	316	82	29	20
3.50	0	0	1	1	1	2	2	2	2	3	3	4	4	5	5	6	7	9	16	46	300	180	34	22
4.00	0	0	0	0	0	1	1	1	1	1	2	2	2	3	3	3	4	6	9	16	53	286	41	24

Table 8. --Tabular discharges for type-II storm distribution (csm/in)--Continued   Sheet 2 of 5

TIME OF CONCENTRATION = 0.3 hours
Hydrograph Time in Hours

Tt	11.0	11.5	11.7	11.8	11.9	12.0	12.1	12.2	12.3	12.4	12.5	12.6	12.7	12.8	12.9	13.0	13.2	13.5	14.0	14.5	15.0	16.0	18.0	20.0
0	21	43	141	324	586	658	535	372	251	184	148	124	102	86	77	71	61	51	41	34	30	24	18	14
0.25	17	31	43	67	134	279	461	559	530	428	318	234	179	143	116	97	76	59	45	37	32	25	18	15
0.50	13	22	29	34	42	65	124	238	378	479	499	447	363	281	216	168	110	74	51	41	34	26	19	15
0.75	10	17	21	24	27	32	41	63	114	203	316	413	457	443	389	319	198	105	60	45	37	28	20	15
1.00	8	13	16	18	20	23	26	31	40	60	103	176	269	358	415	426	344	182	77	51	41	30	20	16
1.50	5	8	10	11	12	13	15	16	18	21	24	28	36	52	82	132	272	382	192	81	52	34	22	17
2.00	3	5	6	7	8	8	9	10	11	12	14	15	17	19	21	25	44	151	351	198	85	41	24	18
2.50	1	3	4	4	5	5	6	6	7	8	8	9	10	11	12	14	17	28	162	328	200	54	27	19
3.00	0	1	2	2	3	3	3	4	4	5	5	6	6	7	8	9	10	14	33	169	309	94	30	20
3.50	0	0	1	1	1	1	2	2	2	3	3	3	4	4	5	5	6	9	14	38	172	294	35	22
4.00	0	0	0	0	0	0	1	1	1	1	3	2	2	2	3	3	4	5	9	15	43	261	42	24

TIME OF CONCENTRATION = 0.4 hours
Hydrograph Time in Hours

Tt	11.0	11.5	11.7	11.8	11.9	12.0	12.1	12.2	12.3	12.4	12.5	12.6	12.7	12.8	12.9	13.0	13.2	13.5	14.0	14.5	15.0	16.0	18.0	20.0
0	20	39	103	224	419	558	575	451	331	247	190	155	127	105	90	80	66	53	42	35	30	24	18	14
0.25	15	28	38	54	98	196	343	467	508	464	380	295	228	180	145	119	87	64	47	38	32	26	19	15
0.50	12	20	26	30	37	53	92	172	286	395	462	453	402	332	266	211	137	84	54	42	35	27	19	15
0.75	10	16	19	22	25	29	36	51	85	150	242	338	407	429	406	356	281	128	65	47	38	29	20	16
1.00	8	12	15	17	19	21	24	28	34	49	78	132	208	292	362	403	368	220	88	55	42	30	21	16
1.50	5	8	9	10	11	12	14	15	17	19	22	25	31	43	65	102	220	365	224	93	56	35	22	17
2.00	3	5	6	6	7	8	9	9	10	11	13	14	16	17	20	23	37	119	338	225	99	43	24	18
2.50	1	3	3	4	4	5	5	6	6	7	8	9	10	11	12	13	16	25	132	317	225	98	27	19
3.00	0	1	2	2	3	3	3	3	4	4	5	5	6	7	7	8	10	13	28	140	300	107	31	21
3.50	0	0	1	1	2	2	2	2	2	2	3	3	3	4	4	5	6	8	13	32	146	286	36	22
4.00	0	0	0	0	0	0	1	1	1	1	1	1	2	2	2	3	3	5	8	14	36	273	44	24

Table 8.--Tabular discharges for type-II storm distribution (csm/in)--Continued   Sheet 3 of 5

TIME OF CONCENTRATION = 0.5 hours

Hydrograph Time in Hours

T_t	11.0	11.5	11.7	11.8	11.9	12.0	12.1	12.2	12.3	12.4	12.5	12.6	12.7	12.8	12.9	13.0	13.2	13.5	14.0	14.5	15.0	16.0	18.0	20.0
0	18	36	80	166	301	433	496	474	395	309	242	194	158	130	109	94	75	57	43	36	31	25	18	15
0.25	15	26	37	52	94	172	277	372	425	424	383	326	270	221	182	150	107	73	49	39	33	26	19	15
0.50	12	20	25	30	38	58	101	169	252	327	374	385	366	329	285	241	169	103	59	44	36	27	19	16
0.75	9	15	19	22	25	30	41	63	103	162	229	292	335	354	348	325	255	157	77	50	39	29	20	16
1.00	7	12	15	17	19	21	25	31	43	66	103	153	210	264	304	327	317	231	109	61	44	31	21	16
1.50	5	8	9	10	11	12	14	15	17	20	24	31	43	63	92	129	214	295	224	115	65	36	23	17
2.00	3	5	6	6	7	8	9	10	11	12	13	14	16	19	23	30	58	143	271	216	120	46	25	18
2.50	1	3	3	4	5	5	5	6	7	7	8	9	10	11	12	14	18	39	150	253	209	71	28	19
3.00	0	1	2	2	2	3	3	4	4	4	5	5	6	7	7	8	10	15	48	154	239	126	32	21
3.50	0	0	1	1	1	1	2	2	2	2	3	3	4	4	5	5	6	8	16	56	155	227	38	23
4.00	0	0	0	0	0	1	1	1	1	1	1	2	2	2	3	3	4	5	9	19	63	217	52	25

TIME OF CONCENTRATION = 0.75 hours

Hydrograph Time in Hours

T_t	11.0	11.5	11.7	11.8	11.9	12.0	12.1	12.2	12.3	12.4	12.5	12.6	12.7	12.8	12.9	13.0	13.2	13.5	14.0	14.5	15.0	16.0	18.0	20.0
0	15	29	57	98	163	248	329	375	388	369	325	276	232	195	165	142	107	76	51	39	33	26	19	15
0.25	12	21	29	39	61	100	158	227	291	336	355	348	321	285	247	212	156	103	62	44	36	27	19	15
0.50	10	16	21	24	29	41	63	100	150	208	263	305	327	329	314	288	226	147	79	52	40	29	20	16
0.75	8	13	16	18	20	24	30	43	65	98	142	192	239	278	303	311	286	208	107	63	45	31	21	16
1.00	6	10	13	14	15	17	20	24	31	44	65	95	134	177	220	256	294	263	149	81	53	33	21	16
1.50	4	6	8	9	10	11	12	13	14	16	19	23	31	42	60	83	147	269	208	152	85	40	23	17
2.00	2	4	5	5	6	7	7	8	9	10	11	12	14	16	18	23	39	97	251	235	153	56	26	19
2.50	1	2	3	3	4	4	4	5	5	6	7	7	8	9	10	11	15	28	107	218	236	91	29	20
3.00	0	0	1	2	2	2	2	3	3	4	4	5	5	6	6	7	8	12	33	113	225	153	34	22
3.50	0	0	0	1	1	1	1	1	2	2	2	3	3	3	4	4	5	7	13	39	117	215	44	24
4.00	0	0	0	0	0	0	0	1	1	1	1	1	1	2	2	2	3	4	7	15	45	207	63	26

Table 8.--Tabular discharges for type-II storm distribution (csm/in)--Continued   Sheet 4 of 5

TIME OF CONCENTRATION = 1.0 hours

Hydrograph Time in Hours

Tt	11.0	11.5	11.7	11.8	11.9	12.0	12.1	12.2	12.3	12.4	12.5	12.6	12.7	12.8	12.9	13.0	13.2	13.5	14.0	14.5	15.0	16.0	18.0	20.0
0	13	24	45	66	107	155	211	258	301	313	316	301	277	247	217	188	146	102	64	46	36	27	19	15
0.25	10	18	24	32	45	68	102	146	193	238	272	293	299	293	275	252	200	139	81	54	41	29	20	16
0.50	8	14	17	20	24	32	46	68	99	136	178	219	251	274	284	283	254	187	105	65	47	31	21	16
0.75	7	11	13	15	17	20	25	33	46	67	94	128	165	202	233	256	273	236	140	82	55	33	21	16
1.00	5	9	11	12	13	15	17	20	25	33	46	65	90	121	154	187	240	262	183	107	66	37	22	17
1.50	3	5	7	7	8	9	10	11	12	14	16	19	24	31	43	58	103	185	244	181	110	48	24	18
2.00	2	3	4	4	5	6	6	7	8	8	9	10	11	13	15	18	29	69	182	230	178	70	27	19
2.50	1	2	2	3	3	3	4	4	5	5	6	6	7	8	9	10	12	21	77	178	219	114	31	21
3.00	0	1	1	1	1	2	2	2	3	3	3	4	4	5	5	6	7	10	25	83	210	172	39	22
3.50	0	0	0	0	1	1	1	1	1	2	1	2	2	3	3	3	4	6	11	29	88	202	52	25
4.00	0	0	0	0	0	0	0	0	1	1	1	1	1	1	2	2	2	4	6	12	33	195	77	28

TIME OF CONCENTRATION = 1.25 hours

Hydrograph Time in Hours

Tt	11.0	11.5	11.7	11.8	11.9	12.0	12.1	12.2	12.3	12.4	12.5	12.6	12.7	12.8	12.9	13.0	13.2	13.5	14.0	14.5	15.0	16.0	18.0	20.0
0	11	21	37	51	79	107	147	187	219	249	264	271	267	256	241	219	177	128	81	56	42	29	20	16
0.25	9	15	21	27	36	53	74	103	137	172	205	231	249	259	259	253	223	167	102	67	48	31	21	16
0.50	7	12	15	17	21	27	37	51	72	98	128	160	190	216	235	247	251	209	130	82	56	34	21	16
0.75	6	9	12	13	15	17	21	27	36	50	69	93	120	149	177	202	235	242	165	103	67	38	22	17
1.00	4	7	9	10	11	13	14	17	21	27	36	49	66	88	113	139	190	235	200	130	83	43	23	17
1.50	3	5	6	6	7	8	8	9	10	12	14	16	20	25	33	44	76	142	223	195	131	58	26	18
2.00	1	3	3	4	4	5	5	6	6	7	8	9	10	11	13	15	24	52	143	212	109	66	29	20
2.50	1	2	2	2	3	3	3	3	4	4	5	5	6	7	7	8	10	17	58	143	201	132	35	21
3.00	0	1	1	1	1	1	2	2	2	2	3	3	3	4	4	5	6	9	20	64	143	196	45	23
3.50	0	0	0	0	0	0	0	1	1	1	1	2	2	2	2	3	4	5	9	23	68	190	62	26
4.00	0	0	0	0	0	0	0	0	0	0	1	1	1	1	1	1	2	3	5	10	26	104	91	30

42

Table 8. —Tabular discharges for type-II storm distribution (csm/in)—Continued   Sheet 5 of 5

TIME OF CONCENTRATION = 1.5 hours

Hydrograph Time in Hours

z_t	11.0	11.5	11.7	11.8	11.9	12.0	12.1	12.2	12.3	12.4	12.5	12.6	12.7	12.8	12.9	13.0	13.2	13.5	14.0	14.5	15.0	16.0	18.0	20.0
0	10	18	31	42	57	81	105	133	164	192	209	227	235	236	236	225	201	153	99	68	50	32	20	16
0.25	8	13	17	22	30	41	57	76	99	125	153	178	199	215	225	230	224	188	122	82	55	36	21	16
0.50	6	10	13	15	18	22	30	40	54	72	94	118	143	167	188	204	224	214	152	99	68	39	22	17
0.75	5	8	10	11	13	15	18	22	29	39	52	69	89	111	134	157	194	219	182	122	82	44	23	17
1.00	4	6	8	9	10	11	12	14	17	22	29	38	50	66	84	105	148	198	214	150	100	50	24	18
1.50	2	4	5	6	7	7	7	8	9	10	12	14	17	21	26	34	58	109	191	204	149	70	28	19
2.00	1	2	3	3	4	4	4	5	5	6	7	8	8	10	11	13	19	40	112	184	197	102	33	20
2.50	0	1	1	2	2	2	3	3	3	4	4	5	5	6	6	7	9	14	45	114	190	147	40	22
3.00	0	0	1	1	1	1	1	1	2	2	2	3	3	3	4	4	5	7	16	49	115	184	53	25
3.50	0	0	0	0	0	0	1	1	1	1	1	1	2	2	2	2	3	4	8	18	53	178	74	28
4.00	0	0	0	0	0	0	0	0	0	0	0	1	1	1	1	1	2	2	4	8	21	174	105	34

TIME OF CONCENTRATION = 2.0 hours

Hydrograph Time in Hours

z_t	11.0	11.5	11.7	11.8	11.9	12.0	12.1	12.2	12.3	12.4	12.5	12.6	12.7	12.8	12.9	13.0	13.2	13.5	14.0	14.5	15.0	16.0	18.0	20.0
0	7	14	22	30	38	49	64	80	95	114	133	152	165	175	184	192	190	176	129	93	68	41	23	17
0.25	6	10	13	17	22	28	37	47	61	75	91	108	126	143	157	168	185	189	153	109	79	46	24	17
0.50	5	8	10	11	13	17	21	27	35	45	57	71	86	103	119	135	162	186	172	129	92	52	26	18
0.75	4	6	8	8	10	11	13	16	21	26	34	43	55	67	82	97	129	166	183	149	109	59	27	18
1.00	3	5	6	7	7	8	9	11	13	16	20	26	33	42	52	64	92	136	180	167	127	68	29	19
1.50	1	3	3	4	4	5	5	6	7	8	9	10	12	15	18	23	37	68	135	175	163	93	34	21
2.00	1	1	2	2	3	3	3	4	4	5	5	6	6	7	8	10	14	26	71	133	170	127	42	23
2.50	0	1	1	1	1	1	2	2	2	3	3	3	4	4	5	5	7	11	29	74	132	166	53	26
3.00	0	0	0	0	1	1	1	1	1	1	2	2	2	2	3	3	4	5	12	32	76	162	71	30
3.50	0	0	0	0	0	0	0	0	1	1	1	1	1	1	1	2	2	3	6	13	35	158	95	35
4.00	0	0	0	0	0	0	0	0	0	0	0	0	0	1	1	1	1	2	3	6	14	80	155	43

Table 8. A table segment is given for selected times-of-concentration (t_c): 0.1, 0.2, 0.3, 0.4, 0.5, 0.75, 1.0, 1.25, 1.5, and 2.0 hours. For values other than these select values, the closest value can be used; additional precision can be achieved through interpolation.

Each table segment is further subdivided by the total travel time (T_t) from the subwatershed outlet to the design point. For each t_c and T_t discharge rates are given in cfs/square mile/inch of runoff for hydrograph times (i.e., the time from the beginning of precipitation) ranging from 11.0 hours to 20.0 hours in various time increments.

Procedure used in solving problems with the tabular discharge hydrograph method is to segment the watershed into appropriate subareas and identify the necessary input for each subarea and channel reach. The hydrograph at the design point due to runoff from any subarea is determined by entering Table 8 for the subarea t_c and the total travel time from the outlet of the subarea to the design point. The total hydrograph is determined by summing the subarea hydrographs. The solution procedure is best illustrated with examples.

Example 10-1

A 1.1 square mile watershed is subdivided into two subareas. The upper portion of the watershed has a drainage area of 0.6 mi^2, a time-of-concentration of 2.0 hours, and a runoff curve number of 70. The lower portion has a drainage area of 0.5 mi^2, a time-of-concentration of 1.5 hours, and a runoff curve number of 75. The travel time from the outlet of the upper portion of the watershed through the channel in the lower portion of the watershed is 1.0 hours. Assume a precipitation of 7 inches.

The hydrograph computations are given in Table 9. The runoff volumes, which are obtained from Fig. 5, are 3.6 inches and 4.1 inches for the upper and lower portions of the watershed, respectively. The hydrograph computations for the upper portion (i.e., subarea 1) yield the hydrograph from the precipitation excess for subarea 1 as estimated for the outlet from subarea 2 (i.e., the hydrograph is routed through the reach in subarea 2). The computations show the unit discharges (csm/in) multiplied by the product of the drainage area (mi^2) and the runoff volume (in). The total flows indicate that the peak discharge of 661 cfs for the total watershed occurs at the hydrograph time of 13.2 hours. The peak from the upper and lower portions of the watershed occurred at 14.0 and 12.8 hours, respectively.

Example 10-2

The watershed shown in Fig. 14 was subdivided into 4 areas, each having similar land use and soil type. The pertinent characteristics are given in Table 10. The peak discharge as measured at the outfall of the total watershed but resulting from rainfall on the upstream subarea occurred at a hydrograph time of 15.0 hours while the peak discharge from subarea 4 occurred at time 12.6 hours. The peak discharge of 574 cfs for the entire watershed occurred at 13.5 hours (Table 11).

TABLE 9. Hydrograph Computations

Hydrograph Time (hours)

Subarea	12.8	13.0	13.2	13.5	14.0
1	42(3.6)(0.6)=91	64(3.6)(0.6)=138	92(3.6)(0.6)=199	136(3.6)(0.6)=294	180(3.6)(0.6)=389
2	236(4.1)(0.5)=484	225(4.1)(0.5)=461	201(4.1)(0.5)=412	153(4.1)(0.5)=314	99(4.1)(0.5)=203
Total	575 cfs	599 cfs	611 cfs	608 cfs	592 cfs

TABLE 10. Watershed and Runoff Characteristics for P = 7 inches

Subarea	Drainage Area (A) (mi^2)	t_c (hrs)	Runoff CN	Runoff Q (inches)	Reach Travel Time (hrs)	AQ
1	0.40	2.0	67	3.3	-	1.32
2	0.25	1.5	71	3.7	0.75	0.92
3	0.20	1.0	75	4.1	0.25	0.82
4	0.30	1.25	81	4.8	1.00	1.44

TABLE 11. Computation of Composite Flood Hydrograph

Hydrograph Time (hours)

Subarea	12.6	12.8	13.0	13.2	13.5	14.0	14.5	15.0
1	8	9	13	18	34	94	176	224
2	24	40	64	95	141	186	163	115
3	53	99	153	197	215	150	88	54
4	390	369	315	255	184	117	81	60
TOTAL	475 cfs	517 cfs	545 cfs	565 cfs	574 cfs	547 cfs	508 cfs	453 cfs

45

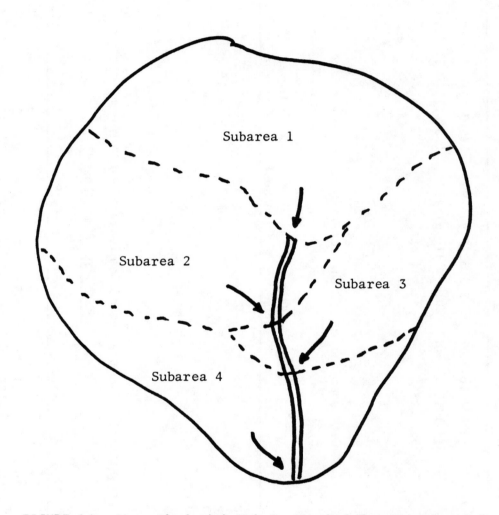

FIGURE 14.   Watershed with Tabular Method for Example 10-2

SECTION 11

THE SCS UNIT HYDROGRAPHS

A hydrograph is a graph of the discharge rate, which passes a particular point, versus time. A hydrograph reflects both precipitation and watershed characteristics, as well as geologic factors. A hydrograph can be separated into three segments: the rising limb, the crest segment, and the recession. The shape of the rising limb is especially sensitive to rainfall characteristics while the shape of the recession is more sensitive to geologic characteristics and watershed slope. The crest segment is sensitive to both watershed and rainfall characteristics.

A total hydrograph consists of both surface, or direct, runoff and baseflow. There are numerous methods for separating baseflow from the total hydrograph. Quite often, the methods identify the inflection point on the recession limb of the hydrograph as the point at which direct runoff ends. However, recognizing that there is no single correct method, it is not unreasonable to assume that the baseflow is constant over the duration of the storm.

Unit Hydrograph: Definition

A unit hydrograph is a special case of the flood hydrograph. Specifically, a unit hydrograph is the hydrograph that results from one inch of precipitation excess generated uniformly over the watershed at a uniform rate during a specified period of time. There are five important concepts in this definition that warrant emphasis. First, the runoff occurs from precipitation excess, which can be defined as the difference between precipitation and losses, which includes interception, depression storage, and infiltrated water that does not appear as direct runoff. Second, the volume of runoff is one inch, which is the same as the volume of precipitation excess. Third, the excess is applied at a constant rate (i.e., uniform rate). Fourth, the excess is applied with a uniform spatial distribution. Fifth, the intensity of the rainfall excess is constant over a specified period of time, which is called the duration.

There are several types of unit hydrographs. The ones of interest herein are the dimensionless and D-hour unit hydrographs. A D-hour unit hydrograph is a unit hydrograph in which the duration of excess is D-hours. A dimensionless unit hydrograph is a hydrograph whose ordinates are given as ratios of the peak discharge (q_p) and whose time axis is measured as a ratio of the time-to-peak (t_p); that is, it is a graph of q/q_p versus t/t_p, in which q is the discharge at any time t.

The SCS Dimensionless Unit Hydrographs

The SCS methods use dimensionless unit hydrographs that are based on an extensive analysis of measured data. Unit hydrographs were evaluated for a large number of actual watersheds and then made dimensionless. An average

47

of these dimensionless unit hydrograph (UH) was developed. The time base of the dimensionless UH was approximately 5 times the time-to-peak and approximately 3/8 of the total volume occurred before the time-to-peak; the inflection point on the recession limb occurs at approximately 1.7 times the time-to-peak, and the UH had a curvilinear shape. The average dimensionless UH is shown in Fig. 15 and the discharge ratios for selected values of the time ratio are given in Table 12.

The curvilinear unit hydrograph can be approximated by a triangular UH that has similar characteristics; Fig. 16 shows a comparison of the two unit hydrographs. While the time base of the triangular UH is only 8/3 of the time-to-peak (compared to 5 for the curvilinear UH), the area under the rising limb of the two UH's are the same (i.e., 37.5%).

The Peak Discharge of the Unit Hydrograph

The area under the unit hydrograph equals the volume of direct runoff Q, which was estimated by Eq. 7:

$$Q = \frac{1}{2} q_p (t_p + t_r) \tag{13}$$

in which t_p and t_r are the time-to-peak and the recession time, respectively; and q_p is the peak discharge. Solving Eq. 13 for q_p and rearranging yields:

$$q_p = \frac{Q}{t_p} \left[\frac{2}{1 + t_r/t_p} \right] \tag{14}$$

Letting K replace the contents within the brackets yields:

$$q_p = \frac{KQ}{t_p} \tag{15}$$

In order to have q_p in cfs, t_p in hours, and Q in inches, it is necessary to **multiply Eq. 15 by the area in square miles and the constant 645.3; also** because $t_r = 1.67 \, t_p$, Eq. 15 is:

$$q_p = \frac{484 \, AQ}{t_p} \tag{16}$$

The constant 484 reflects a unit hydrograph that has 3/8 of its area under the rising limb. For mountainous watersheds the fraction could be expected to be greater and, therefore, the constant of Eq. 16 may be near 600. For flat, swampy areas the constant may be on the order of 300.

The time-to-peak of Eq. 16 can be expressed in terms of the duration of unit precipitation excess and the time of concentration. Fig. 16 provides the following two relationships:

$$t_c + D = 1.7 \, t_p \tag{17}$$

and

48

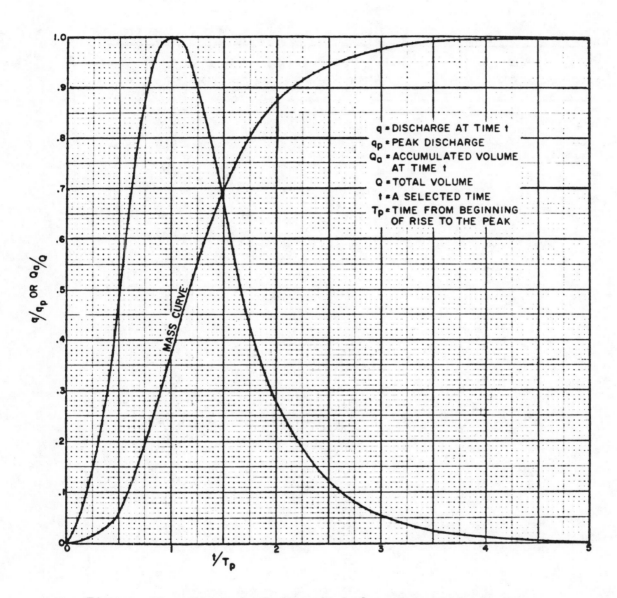

Figure 15. Dimensionless unit hydrograph and mass curve

Table 12.   Ratios for dimensionless unit hydrograph
and mass curve.

Time Ratios (t/T_p)	Discharge Ratios (q/q_p)	Mass Curve Ratios (Qa/Q)
0	.000	.000
.1	.030	.001
.2	.100	.006
.3	.190	.012
.4	.310	.035
.5	.470	.065
.6	.660	.107
.7	.820	.163
.8	.930	.228
.9	.990	.300
1.0	1.000	.375
1.1	.990	.450
1.2	.930	.522
1.3	.860	.589
1.4	.780	.650
1.5	.680	.700
1.6	.560	.751
1.7	.460	.790
1.8	.390	.822
1.9	.330	.849
2.0	.280	.871
2.2	.207	.908
2.4	.147	.934
2.6	.107	.953
2.8	.077	.967
3.0	.055	.977
3.2	.040	.984
3.4	.029	.989
3.6	.021	.993
3.8	.015	.995
4.0	.011	.997
4.5	.005	.999
5.0	.000	1.000

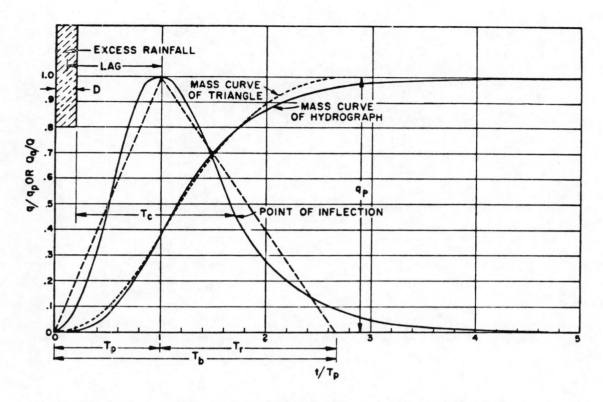

Figure 16. Dimensionless curvilinear unit hydrograph and equivalent triangular hydrograph

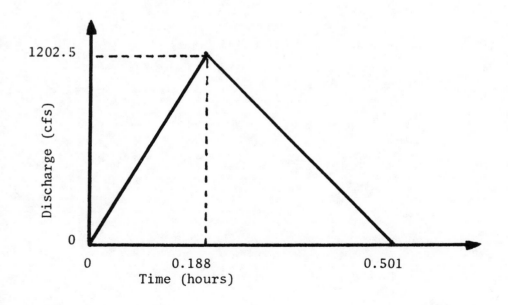

FIGURE 17. The Triangular Unit Hydrograph for Example 11-1

$$\frac{D}{2} + 0.6 \, t_c = t_p \qquad (18)$$

Solving Eqs. 17 and 18 for D yields:

$$D = 0.133 \, t_c \qquad (19)$$

Therefore, t_p can be expressed in terms of t_c:

$$t_p = \frac{D}{2} + 0.6 \, t_c = \frac{2}{3} \, t_c \qquad (20)$$

Expressing Eq. 16 in terms of t_c, rather than t_p, yields:

$$q_p = \frac{726 \, AQ}{t_c} \qquad (21)$$

Example 11-1

Determine the triangular UH for a 300 acre watershed that has been commercially developed. The flow length is 1500 feet, the slope is 3 percent, and the soil is of group B.

For commercial land use and soil group B, the watershed CN is 92. The watershed lag from Fig. 7 is 0.17 hours; therefore, the t_c is 0.283 hours. For 1 inch of precipitation excess, Eq. 21 provides a peak discharge of:

$$q_p = \frac{726 \, (300 \text{ acres})(1 \text{ inch})}{(640 \text{ acres/sq. mile})(0.283 \text{ hrs})} = 1202 \text{ cfs} \qquad (22)$$

The time-to-peak is:

$$t_p = \frac{2}{3} \, t_c = 0.188 \text{ hour} \qquad (23)$$

and the time base of the UH is:

$$t_b = \frac{8}{3} \, t_p = 0.501 \text{ hours} \qquad (24)$$

The resulting triangular UH is shown in Fig. 17.

SECTION 12

CONVOLUTION

The dimensionless unit hydrograph of Fig. 15 is made dimensional by computing the peak discharge and time-to-peak using Eqs. 21 and 20, respectively.
The unit hydrograph results from one inch of precipitation distributed uniformly over the time period D. Thus, the unit hydrograph does not represent
either the total runoff volume or the design hydrograph. To compute the
latter, it is necessary to use the unit hydrograph to translate the time
distribution of precipitation excess into a runoff hydrograph.

The process of translating precipitation excess into a runoff hydrograph is called convolution. Analytically speaking, convolution is referred to as the theory of linear superpositioning. Conceptually, it is a process of multiplication, translation with time, and addition. That is, the
first burst of precipitation excess of duration D is multiplied by the ordinates of the UH. The UH is then translated a time length of D, and the next
burst of rainfall excess is multiplied by the UH. After the UH has been
translated for all bursts of excess of duration D, the results of the multi-
plications are summed for each time interval. This process of multiplication, translation, and addition is the means of deriving a design hydrograph
from the precipitation excess and the UH.

The convolution process is best introduced using some very simple examples that illustrate the multiplication-translation-addition operation.
First, consider a burst of excess of one inch that occurs over a period D.
Assuming a UH that consists of two ordinates, 0.4 and 0.6, the runoff is
computed by multiplying the rainfall excess burst by the UH; this is best
presented graphically as follows:

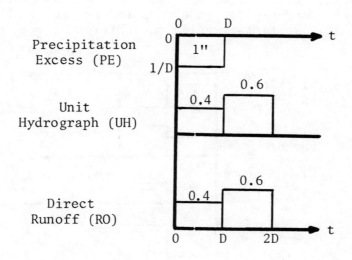

53

It is important to note that the volume of runoff equals the volume of excess, which in this case is one inch.

If 2 inches of precipitation excess occur over a period of D, then the runoff volume will be 2 inches. Using the same UH as the previous example, the resulting runoff hydrograph is as follows:

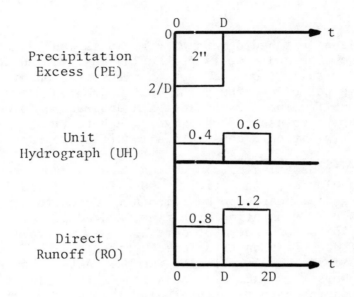

In both this example and the previous example computation of the runoff hydrograph consisted solely of multiplication; the translation and addition parts of the convolution process were not necessary because the precipitation occurred over a time interval D.

To illustrate the multiplication-translation-addition operation, consider 2 inches of precipitation excess that occurs over a period 2D. In this case, the runoff will have a volume of two inches, but the time distribution of runoff will differ from that of the previous problem because the time distribution of precipitation excess is different. The following diagram shows the multiplication-translation-addition operation:

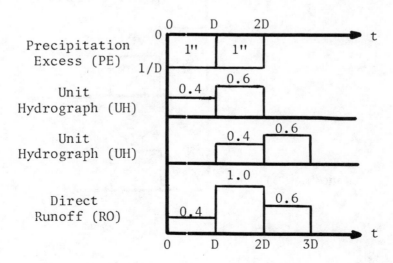

In this case, the time base of the runoff hydrograph is 3 time units long (i.e., 3D).   In general, the time base of the runoff (t_{bRO}) is given by:

$$t_{bRO} = t_{bPE} + t_{bUH} - 1 \qquad (25)$$

in which t_{bPE} and t_{bUH} are the time bases of the precipitation excess and unit hydrograph, respectively.   For the above example both t_{bPE} and t_{bUH} equal two, and therefore, according to Eq. 25 t_{bRO} equals three units.

One more simple example should illustrate the convolution process.   The volume of precipitation excess equals 3 inches, with 2 inches occurring in the first time unit.   The following diagram shows the computation of the runoff hydrograph:

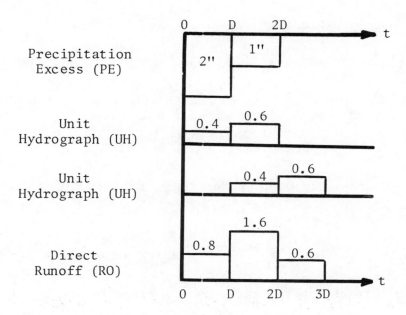

In this case, the second ordinate of the runoff hydrograph is the sum of 2 inches times the second ordinate of the UH and one inch times the first ordinate of the translated UH:

$$2(0.6) + 1(0.4) = 1.6$$

The convolution process is performed in the TR-20 subrouting RUNOFF.   In this case, the distribution of precipitation excess is derived from the distribution of the precipitation of the type I or type II storms.   The derivation of the unit hydrograph follows the procedure outlined in a previous section.   The convolution process outlined here is then applied to develop the surface runoff hydrograph.

Example 12-1

Develop the runoff hydrograph for a 400 acre watershed using the following distribution of precipitation:

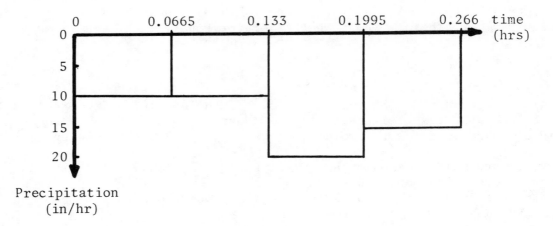

The volume of precipitation is:

$$(10+10+20+15) \ \frac{\text{inches}}{\text{hour}} \ \text{x} \ 0.0665 \ \text{hours} = 3.6575 \ \text{inches}$$

Assume the runoff curve number is 90 and the time-of-concentration is 0.5 hours. Using Eq. 20 the routing increment D is given by

$$D = 0.133 \ t_c = 0.0665 \ \text{hour}$$

The following procedure is used in deriving the runoff hydrograph.

Step 1. Develop the triangular UH

$$t_p = \frac{D}{2} + 0.6 \ t_c = \frac{0.0665}{2} + 0.6(0.5) = 0.333 \ \text{hour}$$

For one inch of direct runoff, the peak discharge is:

$$q_p = \frac{484 \ AQ}{t_p} = \frac{484(400/640)(1)}{0.333} = 908 \ \text{cfs}$$

The time base of the UH is:

$$t_b = \frac{8}{3} \ t_p = 0.889 \ \text{hour}$$

The resulting UH is shown in Fig. 18.

Step 2. Tabulate the UH ordinates using D = 0.665 hrs.

The resulting ordinates are given in column (2) of Table 13.

Step 3. Tabulate the precipitation in time increments of D

The precipitation intensity and volumes are given in columns (3) and (4), respectively, of Table 13. The cumulative precipitation is given in column (5).

TABLE 13. Computation of Incremental Runoff for Example 12-1

(1) t(hrs)	(2) q(cfs)	(3) P(in/hr)	(4) P(in)	(5) Σ P	(6) Q	(7) ΔQ
0	0	0	0	0	0	0
.0665	182	10	0.6650	0.6650	0.13	0.13
.1330	363	10	0.6650	1.3300	0.58	0.45
.1995	545	20	1.3300	2.6600	1.68	1.10
.2660	726	15	0.9975	3.6575	2.61	0.93
.3325	908					
.3990	799					
.4655	691					
.5330	582					
.5995	474					
.6660	365					
.7325	256					
.7990	148					
.8655	39					
.9320	0					

Σ=6078

TABLE 14. Computation of Runoff Hydrograph using Convolution for Example 12-1

t		q
0	0	= 0
0.0665	182(.13)	= 24
0.1330	363(.13) + 182(.45)	= 129
0.1995	545(.13) + 363(.45) + 182(1.1)	= 434
0.2660	726(.13) + 545(.45) + 363(1.1) + 182(.93)	= 908
0.3325	908(.13) + 726(.45) + 545(1.1) + 363(.93)	= 1382
0.3990	799(.13) + 908(.45) + 726(1.1) + 545(.93)	= 1818
0.4655	691(.13) + 799(.45) + 908(1.1) + 726(.93)	= 2123
0.5330	582(.13) + 691(.45) + 799(1.1) + 908(.93)	= 2110
0.5995	474(.13) + 582(.45) + 691(1.1) + 799(.93)	= 1827
0.6660	365(.13) + 474(.45) + 582(1.1) + 691(.93)	= 1544
0.7325	256(.13) + 365(.45) + 474(1.1) + 582(.93)	= 1260
0.7990	148(.13) + 256(.45) + 365(1.1) + 474(.93)	= 977
0.8655	39(.13) + 148(.45) + 256(1.1) + 365(.93)	= 693
0.9320	39(.45) + 148(1.1) + 256(.93)	= 418
0.9985	39(1.1) + 148(.93)	= 181
1.0640	39(.93)	= 36

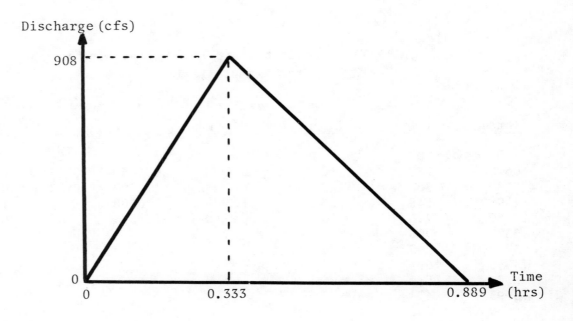

FIGURE 18. Triangular Unit Hydrograph for Example 12-1

Step 4. Compute the accumulated runoff

Given the cumulative precipitation the accumulated runoff volume can be computed for each time increment. The precipitation and curve number are entered into Fig. 5 and the resulting precipitation excess, or runoff, is determined.

Step 5. Compute incremental precipitation excess

The incremented volumes of precipitation excess are determined by subtracting the values for each D. The resulting time distribution is given in column (7) of Table 13.

Step 6. Determine the runoff hydrograph by convolution

The unit hydrograph ordinates of column (2) of Table 13 is convoluted with the precipitation excess of column (7). The computations are shown in Table 14. The resulting runoff hydrograph is also given in Table 14.

Comments. This example illustrates the derivation of the precipitation excess from precipitation, the development of the UH, and the convolution process for the development of the runoff hydrograph. The example used a simplified distribution of precipitation and the triangular UH. In using TR-20 either the type I or the type II distribution of precipitation and the curvilinear UH are used to derive the runoff hydrograph. However, the computational procedure is the same.

SECTION 13

STREAMFLOW ROUTING

There are two types of routing that are important in TR-20: streamflow routing and reservoir routing. Both involve computing the hydrograph at one point in the watershed system given a hydrograph at another point in the system. Reservoir routing is the process of moving an upstream hydrograph through a structure to a point on the downstream side of the structure; the routing process takes into account the storage characteristics of the structure. Streamflow routing is conceptually similar in that a hydrograph is moved from a point upstream to a point downstream, and the channel storage characteristics determine the degree of attenuation of the hydrograph.

The Routing Equation

The current version of TR-20 uses the convex routing method, which is based on the following routing equation:

$$0_{t+\Delta t} = (1-C)0_t + CI_t \tag{26}$$

in which $0_{t+\Delta t}$ = the outflow (i.e., flow at the downstream point) at time $t+\Delta t$; 0_t = the outflow at time t; I_t = the inflow (i.e., flow at the upstream point) at time t; and C = the routing coefficient, where $0 \leq C \leq 1$. The outflow is determined at a time increment of Δt. Rearranging Eq. 26 yields:

$$0_{t+\Delta t} = 0_t - CO_t + CI_t \tag{27}$$

$$= 0_t + C(I_t - 0_t) \tag{28}$$

Solving Eq. 28 for the routing coefficient yields:

$$C = \frac{0_{t+\Delta t} - 0_t}{I_t - 0_t} \tag{29}$$

The relationship between the routing coefficient and routing characteristics can be developed using the notation of Fig. 19. Using the concept of similar triangles yields the following relationship:

$$\frac{0_{t+\Delta t} - 0_t}{\Delta t} = \frac{I_t - 0_t}{K} \tag{30}$$

in which K is the reach travel time. Eq. 30 can be rearranged as:

$$\frac{0_{t+\Delta t} - 0_t}{I_t - 0_t} = \frac{\Delta t}{K} \tag{31}$$

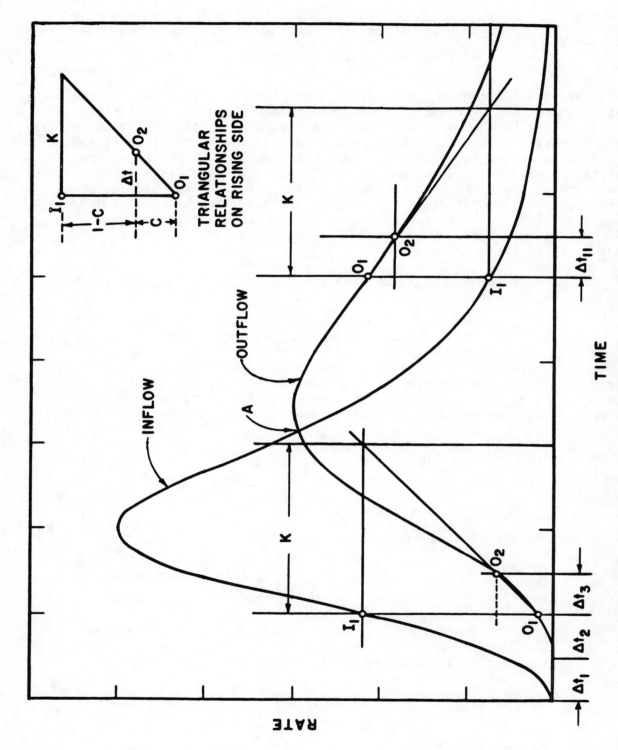

Figure 19. Relationships for Convex method of channel routing.

60

Equating Eqs. 29 and 31 yields the following relationship between the routing coefficient C, the routing interval Δt, and the reach travel time K:

$$C = \frac{\Delta t}{K} \qquad (32)$$

Estimation of the Routing Coefficient

There are several ways that the routing coefficient can be estimated. First, Eq. 32 can be used; the routing interval can be set at approximately one-fifth to one-fourth of the time-to-peak of the unit hydrograph and the reach travel time can be estimated by dividing the reach length by the channel velocity estimated with Manning's equation.

Second, on-site measurements could be used to estimate the reach travel time. Eq. 32 could then be used by setting the routing interval Δt.

Third, studies have shown that the routing coefficient C is approximately twice the value of the routing constant x in the Muskingum method. The constant x reflects the relative importance of inflow and outflow in determining channel storage.

Fourth, an empirical method that is given in NEH-4 uses the steady flow water velocity V with the equation:

$$C = \frac{V}{V + 1.7} \qquad (33)$$

in which V is measured in feet per second. This method is used as a default option in TR-20.

Example 13-1

The objective of streamflow routing is to estimate the flood hydrograph at a downstream point. The solution requires the inflow hydrograph in increments of Δt, the outflow at time t=0, and the routing coefficient. Eq. 26 is applied iteratively to estimate the entire downstream hydrograph.

The inflow hydrograph of Fig. 20 will be used to illustrate the solution procedure. For a velocity of 0.75 fps, Eq. 33 would yield a value of 0.3 for the routing coefficient. By applying Eq. 26 for time increments of 0.5 hours, the outflow hydrograph can be computed. The computations are given in Table 15, and the resulting outflow hydrograph is shown in Fig. 20. The actual peak discharge of the outflow hydrograph does not necessarily occur at 4.0 hours and may be greater than the computed value of 1417 cfs. The actual peak may be obtained either by routing at a smaller routing increment Δt or through curvilinear interpolation of points around the peak value. In any case, Δt should be made small enough so that the peak is accurately estimated and large enough to avoid unnecessarily lengthy calculations. It is recommended that the time increment for a TR-20 computation be no greater than one-fourth of the time-to-peak.

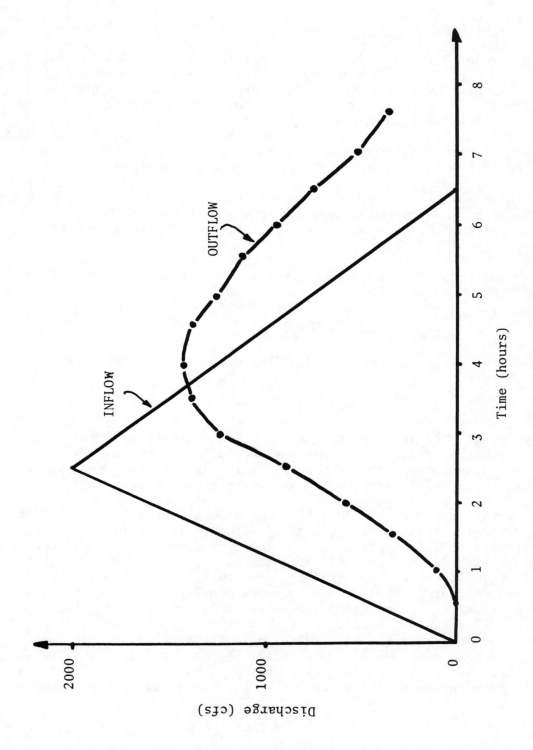

FIGURE 20. Inflow and Routed Outflow Hydrographs for Example 13-1

TABLE 15. Computing of Outflow Hydrograph for Example 13-1

Time (hrs)	Inflow (cfs)	Computation	Outflow (cfs)
0	0	$(1-C)0_t + CI_t$	0
0.5	400	0.7(0) + 0.3(0) =	0
1.0	800	0.7(0) + 0.3(400)	120
1.5	1200	0.7(120) + 0.3(800) =	324
2.0	1600	0.7(324) + 0.3(1200) =	587
2.5	2000	0.7(587) + 0.3(1600) =	891
3.0	1700	0.7(891) + 0.3(2000) =	1224
3.5	1500	0.7(1224) + 0.3(1750) =	1381
4.0	1250	0.7(1381) + 0.3(1500) =	1417
4.5	1000	0.7(1417) + 0.3(1250) =	1367
5.0	750	0.7(1367) + 0.3(1000) =	1257
5.5	500	0.7(1257) + 0.3(750) =	1105
6.0	250	0.7(1105) + 0.3(500) =	923
6.5	0	0.7(923) + 0.3(250) =	721
7.0	0	0.7(721) + 0.3(0) =	504
7.5	0	0.7(504) + 0.3(0) =	353

63

SECTION 14

RESERVOIR ROUTING

The reservoir routing subroutine RESVOR of TR-20 is based on the storage equation:

$$I - 0 = \frac{\Delta S}{\Delta t} \qquad (34)$$

in which I = the input, 0 = the output, Δt = the routing time interval, and ΔS = the change in storage that occurs during the time interval Δt. Both the inflow I and outflow 0 are time varying functions, with I being the inflow hydrograph and 0 being the outflow hydrograph. The inflow hydrograph may be a hydrograph generated with the TR-20 subroutine RUNOFF or a hydrograph reflecting flow that was routed through a reach. While the inflow hydrograph is known, the objective of the reservoir routing subroutine is to compute the outflow hydrograph. The storage equation can be rewritten as:

$$I \Delta t - 0 \Delta t = \Delta S \qquad (35)$$

Figure 19 shows the inflow and outflow hydrographs. If the subscripts 1 and 2 are used to indicate time t and $t+\Delta t$, respectively, the average inflow and average outflow can be used to expand Eq. 35:

$$\frac{1}{2}(I_1 + I_2) \Delta t - \frac{1}{2}(0_1 + 0_2) \Delta t = S_2 - S_1 \qquad (36)$$

While I_1, I_2, 0_1, and S_1 are known at any time t, values for 0_2 and S_2 are unknown. Eq. 36 can be rearranged such that the knowns are placed on one side of the equal sign and the unknowns on the other side:

$$\frac{1}{2}(I_1 + I_2) \Delta t + (S_1 - \frac{1}{2} 0_1 \Delta t) = S_2 + \frac{1}{2} 0_2 \Delta t \qquad (37)$$

Eq. 37 represents one equation with 2 unknowns. In order to find the outflow hydrograph, it is only necessary to compute the outflow-storage relationship, which is easily obtained from site data.

A solution for Eq. 37 can be obtained by deriving the storage-indication curve, which is the relationship between 0 and $(S + \frac{1}{2} 0 \Delta t)$. Given the storage discharge curve, 0 vs. S, the following four step procedure can be used to develop the storage-indication curve:

1. Select a value of 0;

2. Determine the corresponding value of S from the storage-discharge curve;

3. Use the values of S and 0 to compute $(S + \frac{1}{2} 0 \Delta t)$; and

4. Plot 0 vs. $(S + \frac{1}{2} 0 \Delta t)$.

64

These four steps are repeated for a sufficient number of values of 0 to define the storage-indication curve.

The objective of the storage-indication method is to derive the outflow hydrograph. There are five data requirements:

1. The storage-discharge relationship;

2. The storage-indication curve;

3. The inflow-hydrograph;

4. Initial values of the storage and outflow rate; and

5. The routing increment.

The following five step procedure can be used to derive the outflow hydrograph, with the storage-time relationship as a by-product:

1. Determine the average inflow: $\frac{1}{2}(I_1+I_2)\ \Delta t$;

2. Determine $(S_1-\frac{1}{2}\ 0_1\ \Delta t)$;

3. Using Eq. 37 and the values from steps 1 and 2, compute $(S_2+\frac{1}{2}\ 0_2\ \Delta t)$;

4. Using the value computed in step 3, as input, find 0_2 from the storage-indication curve; and

5. Use 0_2 with the storage-discharge relationship, obtain S_2.

These five steps are repeated for the next time increment using I_2, 0_2, and S_2 as the new values of I_1, 0_1, and S_1, respectively. The process is solved iteratively until the outflow hydrograph is computed.

Example 14-1

The storage-discharge relationship is given by:

$$S = 1200\ (0)^{1.035} \tag{38}$$

It is shown in Fig. 21. Using a time increment of 600 seconds values for the storage-indication curve are derived in the following table, and the resulting curve is also shown in Fig. 21:

0 (cfs)	S (ft^3)	S+^{1}/2 0 Δt
5	6350	7850
10	13000	16000
20	26650	32650
30	40550	49550
40	54600	66600

Using both an initial storage and outflow of zero and the inflow hydrograph shown in Fig. 22, Table 16 shows the computations for the outflow hydrograph with a time increment of 10 minutes, or 600 seconds. The resulting outflow hydrograph is shown in Fig. 22.

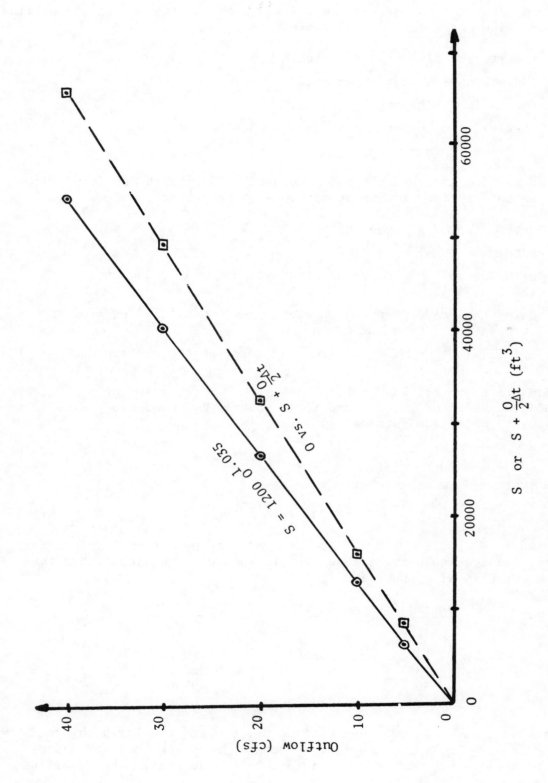

FIGURE 21. The Storage-Discharge and Storage Indication Curves for Example 14-1

TABLE 16. Computation of Reservoir Outflow Hydrograph for Example 14-1

Time (min)	Inflow (cfs)	Average Inflow (cfs)	$0.5(I_1+I_2)\Delta t$	$S-0.5\,O\,\Delta t$	$S+0.5\,O\,\Delta t$	Outflow (O) (cfs)	Storage (S) (ft^3)
0	0			0	4500	0	0
10	15	7.5	4500	2749	17748	2.9	3620
20	35	25.0	15000	11113	30613	11.1	14430
30	30	32.5	19500	19342	35842	18.8	24977
40	25	27.5	16500	22704	34704	21.9	29273
50	15	20.0	12000	21971	27971	21.2	28337
60	5	10.0	6000	17647	19147	17.2	22808
70	0	2.5	1500	12005	12005	11.9	15576
80	0	0	0	7468	7468	7.6	9735
90	0	0	0	4606	4606	4.7	6035
100	0	0	0			3.0	3707

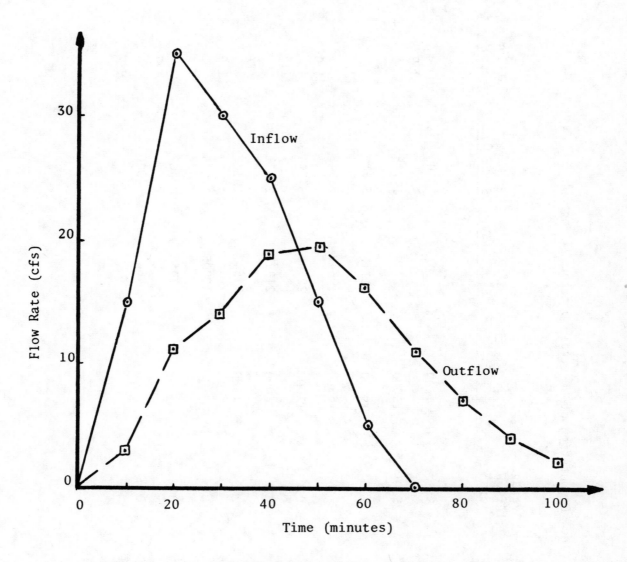

FIGURE 22.   Inflow and Outflow Hydrographs for Example 14-1

68

SECTION 15

HYDROGRAPH SEPARATION

The TR-20 program provides for the separation of a hydrograph into two hydrographs. The DIVERT subroutine is used. The algorithm provides for two hydrograph separation options.

Option 1: The first option is used where a control structure blocks the channel and allows flow down the channel through one or more pipes only. The remaining flow is over a weir at or above the elevation required for full pipe flow into either off-channel storage or a floodway. Operationally, this option separates the inflow hydrograph into two, with outflow hydrograph #1 consisting of all flow below the target discharge; output hydrograph #2 consists of all flow above the target discharge. This option is shown in Fig. 23. The target discharge Q_t is the input parameter.

Option 2: The second option is used where a stream branches into two streams. The inflow hydrograph is divided proportionately to two elevation versus discharge rating tables, both of which must be described in the tabular data. It is important to emphasize that the same datum must be used for the two downstream sections because the surface elevations of the two output hydrographs will be the same.

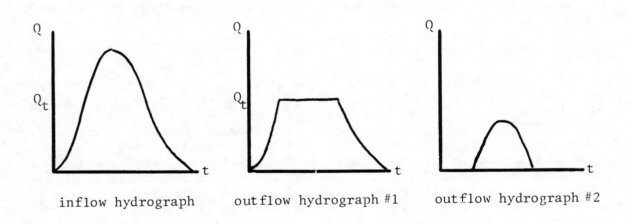

inflow hydrograph   outflow hydrograph #1   outflow hydrograph #2

FIGURE 23. Hydrograph Separation: Option 1

SECTION 16

TR-20

TR-20 is a computerized method for solving hydrologic problems using the concepts outlined in the previous sections. The program was formulated to develop runoff hydrographs, route hydrographs through both channel reaches and reservoirs, and combine or separate hydrographs at confluences. The program is designed to make multiple analyses in a single run so that various alternatives can be evaluated in one pass through the program; this leads to more efficient use of computer time.

Summary of Input Requirements

Even though a computer is used to solve problems, the input data requirements are surprisingly minimal, with the amount of data depending on the complexity of the problem to be solved. If actual rainfall events are not going to be used, the depth of precipitation is the only meteorological input. For each subarea, the drainage area, runoff curve number, and time-of-concentration are required; the antecedent soil moisture condition (i.e., I, II, or III) must also be specified. For each channel reach, the length is required along with the channel cross-section description, which is the elevation, discharge and end area data; while optional, a routing coefficient may also be used as input. If the routing coefficient for Eq. 26 is not given as input, then it will be computed using Eq. 33 and the cross-section data. For each structure it is necessary to describe the outflow characteristics with the elevation-discharge-storage relationship. The time increment for all computations must be specified, and any baseflow in a channel reach must be identified.

Basic Structure of Input Data

Input requirements for a single program run can be separated into three parts: tabular data, standard control, and executive control. The tabular data, which appears first in the data deck consists of the following: 1) the routing coefficient table, which is a tabular statement of Eq. 33; 2) the dimensionless hydrograph table, which is a tabular representation of the curvilinear unit hydrograph of Table 12; 3) one or more cumulative rainfall tables, which provide the depth-time relationship of rainfall input; 4) stream cross-section data, which describes the elevation-discharge-end area relationship; and 5) structure data, which defines the elevation-discharge-storage relationship. The tabular data serves as support data for the problem description.

Following the tabular data, the user must specify the standard control, which consists of cards that establishes the sequence in which runoff hydrographs are determined, routed through reaches and/or structures, and combined at tributary junctions. That is, standard control is the cards that describes the configuration of the watershed and the sequence of runoff computations.

Executive control cards follow the standard control. Executive control controls the computational process. That is, it initiates the execution of the standard control sequence.

The card deck arrangement is shown schematically in Fig. 24.

In addition to tabular data, standard control, and executive control, operations called "modify standard control" are available for solving problems. The modify standard control operations are used when more than one computational sequence is to be tested for a watershed in a single computer run. These operations will be discussed later. However, the modify standard control cards usually follow the executive control cards for the previous problem execution. Additionally, the modify standard control cards are followed by another set of executive control cards, which serve to initiate execution of the problem as described by the computational sequence as altered by the modified standard control.

STANDARD CONTROL

Standard Control Operations

Standard control consists of six subroutine operations:

1. RUNOFF: an instruction to develop an inflow hydrograph;

2. RESVOR: an instruction to route a hydrograph through a structure;

3. REACH: an instruction to route a hydrograph through a channel reach;

4. ADDHYD: an instruction to combine two hydrographs;

5. SAVMOV: an instruction to move a hydrograph from one computer memory storage location to another; and

6. DIVERT: an instruction for a hydrograph to be separated into two parts.

The operations are indicated on the computer card images in columns 2, 4-9, and 11. A 6 is placed in column 2 to indicate standard control. The name of the subroutine operation is placed in columns 4-9. The operation number, which is shown preceding the operation name above (e.g., 1 for RUNOFF, 5 for SAVMOV), is placed in column 11.

Machine Storage

Internal machine storage is used to store the hydrograph ordinates. Machine storage is indicated by numbers from 1 to 7. Columns 19, 21, and 23 are used on standard control cards to indicate machine storage locations. It is important to emphasize that only one hydrograph can occupy any one storage element at a time and that there must be a hydrograph in the storage element from which a subroutine references a storage location. The actual use of the machine storage numbers (i.e., 1 to 7) and the columns (i.e., 19, 21, and 23) will be discussed separately for each subroutine operation.

Output Options

Columns 61, 63, 65, 67, 69, and 71 of standard control cards are used

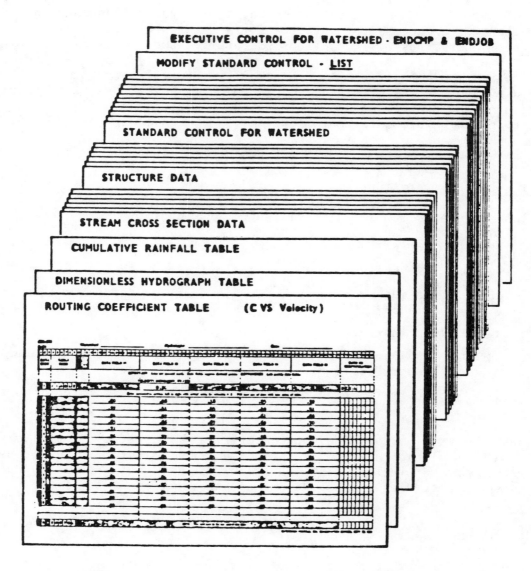

FIGURE 24. Schematic of TR-20 Card Sequence

to specify the output. The individual output options are selected by placing a 1 in the appropriate column. If the column is left blank or a zero is inserted, then the option will not be selected. The following list summarizes the output options:

Output Option	A "1" in Column	Produces the Following Printout
PEAK	61	Peak discharge and corresponding time-of-peak and elevation (maximum stage for a cross-section and maximum storage elevation for a structure).
HYD	63	Discharge hydrograph ordinates.
ELEV	65	Stage hydrograph ordinates (reach elevations for a cross-section and water surface elevation for structures).
VOL	67	Volume of water under the hydrograph in inches-depth, acre-feet, and cfs-hours.
PUNCH	69	The hydrograph and related information is written on a tape that is later used to produce punched cards with a "Read-Discharge Hydrograph" format.
SUM	71	Requests the results of the subroutine be inserted in the summary tables at the end of the job.

Card Identification

Columns 73-80 are available for the user to label the cards. Any alphanumeric characters, including blank spaces, can be used. These spaces are usually used to number the cards, which indicates the sequence. Card identification is important if the program deck gets shuffled by accident.

The card identification columns are available to the user for all cards in the deck and not just for standard control.

Content of Standard Control Cards

In addition to the reserved spaces discussed previously, there are three data fields available for input on each standard control card. The input for each subroutine will be discussed separately.

RUNOFF. The RUNOFF subroutine is intended to develop runoff hydrographs for subareas of a watershed. Input to the subroutine include the drainage area, the CN, and the time of concentration. It uses Eq. 7 in computing the rainfall excess for a given CN and the cumulative rainfall table. Eqs. 20 and 16 are used to compute the time-to-peak and the peak discharge, respectively. The curvilinear dimensionless unit hydrograph is made dimensional and used in convoluting the rainfall excess to develop a runoff hydrograph.

The input for the RUNOFF subroutine is given in Table 17.

TABLE 17. The RUNOFF Card

Column(s)	Content
2	Data Code - Use 6
4-9	Indicator - Use RUNOFF
11	Subroutine Number - Use 1
13-17	X-section (13-15) or Structure (16-17) Number identifying the location in the watershed. Do not insert both a X-section or structure number.
23	Hydrograph Number: Insert number of machine storage element (1 to 7) where hydrograph is stored.
25-36*	Drainage Area (Sq. miles)
37-48*	Runoff curve number **
49-60*	Time-of-Concentration (hours)
61	Output option PEAK
63	Output option HYD
65	Output option ELEV
67	Output option VOL
69	Output option PUNCH
71	Output option SUM
73-80	Card No. Identification: Available for user

* Note: punch decimal point but do NOT include commas to indicate thousands

**Note: Insert the runoff curve number for antecedent moisture condition II; for either condition I or III the machine will make the adjustment in the curve number as instructed on the COMPUT card (column 69).

To illustrate the RUNOFF operation, the following card image would produce a hydrograph for a 12.5 square mile watershed in which the CN was 67 and the time-of-concentration was 3.4 hours. The resulting inflow hydrograph, which represents inflow at cross-section 3, is stored in machine storage 6. The peak discharge is printed.

DATA CODE	SUBROUTINE (OPERATION)		X SECTION STRUCTURE		HYDROGRAPH NUMBER			DATA FIELD #1	DATA FIELD #2	DATA FIELD #3	OUTPUT OPTIONS					CARD NO./ IDENTIFICATION
	NAME	NO.	XSECT NO.	STRUCT NO.	INPUT #1	INPUT #2	OUTPUT				PEAK	HYD	ELEV	VOL	SUM	
6	RUNOFF	1	003				6	12.5	67.0	3.4	1					10

IMPORTANT: Line out unused cards. Data fields require decimal points. KEYPUNCHER: Left justify data fields.

RESVOR. The RESVOR subroutine uses the concepts presented in the section on reservoir routing to route a hydrograph through a structure. The structure number is placed in columns 16-17 and the starting surface elevation in columns 25-36. The machine storage numbers of the inflow and outflow hydrographs are placed in columns 19 and 23, respectively. Table 18 shows the content of a RESVOR card.

The following card image would route the hydrograph in machine storage 6 through the structure numbered 3; the outflow hydrograph would be stored in machine storage unit 7. The surface elevation at time 0 is 15.3 feet.

DATA CODE	SUBROUTINE (OPERATION)		X SECTION STRUCTURE		HYDROGRAPH NUMBER			DATA FIELD #1	DATA FIELD #2	DATA FIELD #3	OUTPUT OPTIONS					CARD NO./ IDENTIFICATION
	NAME	NO.	XSECT NO.	STRUCT NO.	INPUT #1	INPUT #2	OUTPUT				PEAK	HYD	ELEV	VOL	SUM	
6	RESVOR	2		03	6		7	15.3			1					20

IMPORTANT: Line out unused cards. Data fields require decimal points. KEYPUNCHER: Left justify data fields.

REACH. The REACH subroutine is used to route a hydrograph through a channel reach using the method outlined in the section on channel routing. The machine storage element that contains the upstream hydrograph is inserted into column 19; the routed hydrograph is identified by the machine storage element indicated in column 23. The length of the reach, in feet, is inserted in columns 25-36. The routing coefficient of Eq. 27 can be inserted in columns 37-48; if these columns are left blank, then the routing coefficient will be computed using the cross-section data and the routing coefficient table, both of which are described in the section on Tabular Data. The number of routings, which is also optional, is placed in columns 49-60.

The input for the REACH subroutine is given in Table 19.

The following card image would be used to route the hydrograph stored in machine element 7 to cross-section 004; the downstream hydrograph is placed in machine storage element 5. The reach length is 2350 feet. Because columns 37-48 are blank, the routing coefficient will be computed internally using the data for cross-section 004. The peak discharge is printed.

TABLE 18. The RESVOR Card

Column(s)	Content
2	Data Code - Use 6
4-9	Indicator - Use RESVOR
11	Subroutine Number - Use 2
16-17	Number identifying the structure
19	Hydrograph Number: insert number of machine storage element that contains the inflow hydrograph
23	Hydrograph Number: insert number of machine storage element into which the outflow hydrograph is stored
25-36*	Surface Elevation (feet) at time zero
61	Output Option PEAK
63	Output Option HYD
65	Output Option ELEV
67	Output Option VOL
69	Output Option PUNCH
71	Output Option SUM
73-80	Card No. Identification: Available for User

* Note: punch decimal point but do NOT include commas to indicate thousands

TABLE 19.   The REACH Card

Column(s)	Content
2	Data Code - Use 6
4-8	Indicator - Use REACH
11	Subroutine Number - Use 3
13-15	Cross-section Number identification
19	Hydrograph Number: insert number of machine storage element that contains the inflow hydrograph.
23	Hydrograph Number: Insert number of machine storage element into which the outflow hydrograph is stored.
25-36*	Length (feet) of stream reach
37-48**_***	Routing Coefficient for Convex Routing Method.
49-60**	No. of Routings
61	Output Option PEAK
63	Output Option HYD
65	Output Option ELEV
67	Output Option VOL
69	Output Option PUNCH
71	Output Option SUM
73-80	Card No. Identification: Available for user.

* Note: punch decimal point but do NOT include commas to indicate thousands.

**Note: optional

***Note: note the option of specifying a routing coefficient (C). If the steady flow velocity for the routing reach has been precomputed for one reason or other, a corresponding routing coefficient "C" can be inserted. The machine will compute the modified coefficient "C" for the reach routing without searching for a cross section with which to make the computation. Conversely, if a coefficient "C" is not shown and the space is left blank, the machine will compute the routing coefficient from the appropriate cross-section data and C-table.

1 2 3	4 5 6 7 8 9 10 11	12 13	14 15	16 17 18	19	20 21	22 23	24 25 26 27 28 29 30 31 32 33 34 35 36	37 38 39 40 41 42 43 44 45 46 47 48 49 50	51 52 53 54 55 56 57 58 59 60	61 62 63 64 65 66	67 68 69 70 71 72 73 74 75 76 77 78 79 80
DATA CODE	SUBROUTINE (OPERATION)			X SECTION/ STRUCTURE	HYDROGRAPH NUMBER			DATA FIELD #1	DATA FIELD #2	DATA FIELD #3	OUTPUT OPTIONS	CARD NO./ IDENTIFICATION
	NAME	NO.	XSECT NO	STRUCT NO	INPUT #1	INPUT #2	OUT PUT				PRINT PEAK HYD ELEV VOL SUM	

IMPORTANT: Line out unused cards. Data fields require decimal points. KEYPUNCHER: Left justify data fields. Put "1" in space.

| G | REACH | 3 | 04 | | 7 | | 5 | 2350.0 | | | | 30 |

ADDHYD. The ADDHYD subroutine is used to add the two hydrographs that are located in the machine storage elements in columns 19 and 21; the resulting hydrograph is stored in the machine storage element indicated in column 23. The cross-section number where the two hydrographs are added is indicated in columns 13-15. The input for a SAVMOV card image is outlined in Table 20.

SAVMOV. The SAVMOV subroutine is used to move a hydrograph from one machine storage element to another; the respective machine storage elements are indicated in columns 19 and 23. The cross-section or structure number is indicated in columns 13-17. Table 21 summarizes the input requirements for the SAVMOV subroutine.

DIVERT. The DIVERT subroutine is used to separate an upstream hydrograph into hydrographs. The two options available for this subroutine were discussed in the section Hydrograph Separation. The card image input is given in Table 22. Note that for option #2 columns 25-36 will be left blank.

Example 16-1

A 1.5 square mile watershed is subdivided, with the characteristics as follows:

Subarea	Area (mi^2)	CN	t_c (hour)
1	0.8	81	0.5
2	0.7	67	0.75

A detention basin is located at the outlet of the upper subarea, with a surface elevation of 57.7 feet at the start of precipitation. The outflow from the detention basin passes through a 4000 ft. reach, which flows through subarea 2.

A schematic diagram of the watershed is shown in Fig. 25. The standard control card images that are necessary to compute the flood hydrograph for the 1.5 mi^2 watershed are given in Table 23.

Example 16-2

If the channel characteristics of the first 2200 feet in the previous example are different from those in the lower 1800 ft. The problem would have to include a cross-section at the point where the channel characteristics show significant change. The schematic diagram is given in Fig. 26, and the standard control card images are shown in Table 24.

78

TABLE 20.   The ADDHYD Card

Column(s)	Content
2	DATA Code - Use 6
4-9	Indicator - Use ADDHYD
11	Subroutine Number - Use 4
13-15	Cross-section Number indicating where the two hydrographs are added.
19	Hydrograph Number - insert number of machine storage element that contains one of the two inflow hydrographs to be added.
21	Hydrograph Number - insert number of machine storage element that contains the other inflow hydrograph to be added.
23	Hydrograph Number - insert number of machine storage element into which the sum of the two inflow hydrographs is stored.
61	Output Option PEAK
63	Output Option HYD
65	Output Option ELEV
67	Output Option VOL
69	Output Option PUNCH
71	Output option SUM
73-80	Card No. Identification: Available for User

TABLE 21.   The SAVMOV Card

Column(s)	Content
2	Data Code - Usc 6
4-9	Indicator - Use SAVMOV
11	Subroutine Number - Use 5
13-17	Cross-section Number (13-15) or Structure Number (16-17) indicating the reach or structure hydrograph
19	Hydrograph Number: insert number of machine storage element that contains the hydrograph to be moved.
23	Hydrograph Number: insert number of machine storage element into which the hydrograph is to be stored after moving it.
73-80	Card No. Identification: Available for User

TABLE 22. The DIVERT Card

Column(s)	Content
2	Data Code - use 6
4-9	Indicator - use DIVERT
11	Subroutine Number - use 6
13-17	Cross-section Number (13-15) or Structure Number (16-17) for output hydrograph #1.
19	Storage Element Number
21	Storage Element Number for output hydrograph #1
23	Storage Element Number for output hydrograph #2
25-36	The discharge used to separate the hydrograph into two parts (option #1)
37-48	The fraction of the drainage area that goes with output hydrograph #1; the remainder goes with output hydrograph #2
49-60	The cross-section identification number for the output hydrograph #2 (put in decimal form). If left blank, the next sequential cross-section identification number or structure number greater than that used for output hydrograph #1 is assigned to output hydrograph #2
61	Output option PEAK
63	Output option HYD
65	Output option ELEV
67	Output option VOL
69	Output option PUNCH
71	Output option SUM
73-80	Card No. Identification: Available for User

TABLE 23. Standard Control Card Images for Example 16-1

STANDARD CONTROL

Watershed _____ Hydrologist _____ Date _____ Page ____ of ____

IMPORTANT: Line out unused cards. Data fields require decimal points. KEYPUNCHER: Left justify data fields.

Put "1" in space

DATA CODE	SUBROUTINE (OPERATION) NAME	NO.	XSECT NO.	STRUCT. NO.	INPUT #1	INPUT #2	OUTPUT	DATA FIELD #1	DATA FIELD #2	DATA FIELD #3	OUTPUT OPTIONS PRINT — PEAK HYD ELEV VOL RUNOFF / PUNCH / SUM	CARD NO./IDENTIFICATION
6	RUNOFF	1	001		5			0.8	81.0	0.5	1	1
6	RESVOR	2	01		5		6	57.7			1	2
6	RUNOFF	1	002		5			0.7	67.0	0.75	1	3
6	REACH	3	003		6		7	4000.0			1	4
6	ADDHYD	4	003		7	5	6				10001	5
ENDATA												6

NOTE: This card to be inserted at end of all standard control cards.

TABLE 24. Standard Control Card Images for Example 16-2

STANDARD CONTROL

Watershed ____ Hydrologist ____ Date ____

IMPORTANT: Line out unused cards. Data fields require decimal points. KEYPUNCHER: Left justify data fields.

DATA CODE	SUBROUTINE (OPERATION) NAME	NO.	XSECT NO.	STRUCT NO.	INPUT #1	INPUT #2	OUT-PUT	DATA FIELD #1	DATA FIELD #2	DATA FIELD #3	PEAK	HYD	ELEV	VOL	RUNOFF	SUM	CARD NO./ IDENTIFICATION
6	RUNOFF	1	001				5	0.8	81.0	0.5	1						1
6	RESVOR	2	01	5	6			57.7			1						2
6	RUNOFF	1	002				5	0.7	67.0	0.75	1						3
6	REACH	3	003		6		7	2200.0			0						4
6	REACH	3	004		7		6	1800.0			1						5
6	ADDHYD	4	004		6	5	7				1	1	1				6
E N D A T A																	7

OUTPUT OPTIONS — PRINT: Put "1" in space

NOTE: This card is to be used only at end of all standard control cards.

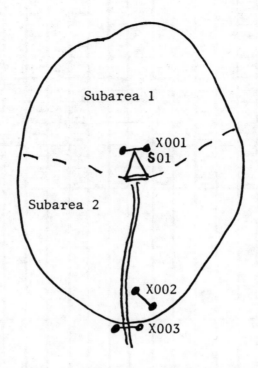

FIGURE 25.   Schematic Diagram of Watershed for Example 16-1

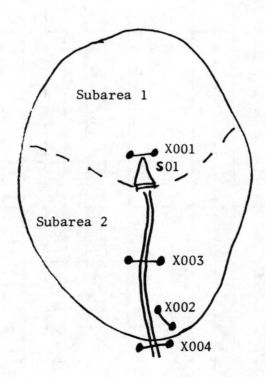

FIGURE 26.   Schematic Diagram of Watershed for Example 16-2

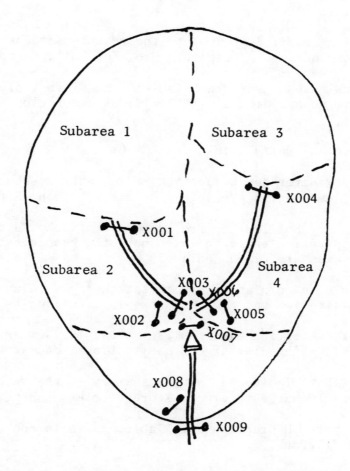

FIGURE 27. Schematic of Watershed for Example 16-3

<u>Example 16-3</u>

A schematic diagram of a 60.8 square mile watershed is shown in Fig. 27. The characteristics of the subareas are given in Table 25 and the channel reaches in Table 26. The reservoir located at the confluence of subareas 2 and 4 has an initial water surface elevation of 20.5 feet. The standard control card images are given in Table 27.

TABULAR DATA

The tabular data, which precedes both the standard and executive control, has six tables:

1. The Routing Coefficient Table: the relationship between the streamflow routing coefficient (C of Eq. 26) and velocity;

2. The Dimensionless Hydrograph Table: the dimensionless curvilinear unit hydrograph ordinates (Table 12) as a function of dimensionless time;

3. Cumulative Rainfall Tables

    a. One-day Watershed Evaluation Storm: the 24-hour dimensionless, cumulative precipitation distribution (Figure 2);

    b. Emergency Spillway or Freeboard Hydrograph: a cumulative precipitation table in which both the precipitation and time axes are dimensionless;

    c. Actual: a cumulative precipitation table in which both the precipitation and time axes are in actual units;

4. Stream Cross-Section Data Tables: a tabular summary of the water surface elevation-discharge-cross sectional end area relationship;

5. Structure Data Table: a tabular summary of the water surface elevation-discharge-reservoir storage relationship; and

6. Read-Discharge Hydrograph Data Table: a table containing an actual hydrograph.

These tables serve as support data for the standard and executive control card images. Not all problems will include all of the tables.

<u>The Routing Coefficient Table</u>

The routing coefficient table gives the tabular relationship between the routing coefficient (C) of Eq. 26 and the channel velocity (V) by Eq. 33. This table is required when a routing coefficient is <u>not</u> inserted on a REACH card image. In some programs, this table is not required because it is contained internally within the program.

Table 28 provides a summary of the input that is required to use the Routing Coefficient Table. The tabular data must be preceded by a CTABLE card and followed by an ENDTBL card. The velocity increment used in the

86

TABLE 25. Watershed Characteristics for Example 16-3

Subarea	Area (mi^2)	Runoff CN	t_c (hrs)
1	10.0	75.0	2.0
2	12.0	68.0	2.7
3	7.8	80.0	1.5
4	11.0	71.0	3.0
5	20.0	72.0	6.1

TABLE 26. Channel Characteristics for Example 16-3

Channel Reach	Length (ft)	Routing Coefficient
003	2000	0.4
006	1800	0.45
009	2700	0.3

TABLE 27. Standard Control Card Images for Example 16-3

STANDARD CONTROL

Watershed _____ Hydrologist _____ Date _____

IMPORTANT: Line out unused cards. Data fields require decimal points. KEYPUNCHER: Left justify data field.

Put "1" in space.

DATA CODE	SUBROUTINE (OPERATION) NAME	NO.	XSECT NO.	STRUCT. NO.	INPUT #1	INPUT #2	OUT-PUT	DATA FIELD #1	DATA FIELD #2	DATA FIELD #3	PEAK	HYD	ELEV	VOL	PUNCH	SUM	CARD NO./ IDENTIFICATION
6	RUNOFF	1	001				5	10.0	75.0	2.0					1		1
6	RUNOFF	1	002				6	12.0	68.0	2.7					1		2
6	REACH	3	003	5			7	2000.0	0.4						1		3
6	ADDHYD	4	003	7	6		5								1		4
6	SAVMOV	5	003	5			1								1		5
6	RUNOFF	1	004				5	7.8	80.0	1.5					1		6
6	RUNOFF	1	005				6	11.0	71.0	3.0					1		7
6	REACH	3	006	5			7	180.	0.45						1		8
6	ADDHYD	4	004	7	6		5								1		9
6	SAVMOV	5	003	1			6								1		10

NOTE: This card is to be used only at end of all standard control cards.

88

TABLE 27. (cont.)

STANDARD CONTROL

Watershed _____   Hydrologist _____   Date _____

DATA CODE (1·2·3)	SUBROUTINE (OPERATION) NAME	NO.	X SECTION/STRUCTURE XSECT NO.	STRUCT NO.	HYDROGRAPH NUMBER INPUT #1	INPUT #2	OUT-PUT	DATA FIELD #1	DATA FIELD #2	DATA FIELD #3	OUTPUT OPTIONS — PRINT PEAK	HYD	ELEV	VOL	PUNCH PUT	SUM	CARD NO./IDENTIFICATION
6	ADDHYD	4	007		5	4	7					1				Put "1" in space.	11
6	RESVOR	2	01	7	5			20.5				1					12
6	RUNOFF	1	008		6			20.0	72.0	6.1		1					13
6	REACH	3	009		5		7	2700.0	0.3			1					14
6	ADDHYD	4	009		4	7	5				0	1	0	1			15
ENDATA																	16

IMPORTANT: Line out unused cards. Data fields require decimal points. KEYPUNCHER: Left justify data fields.

NOTE: This card is to be used only at end of all standard control cards.

89

TABLE 28

ROUTING COEFFICIENT TABLE **

CTABLE Card

Column(s)	Content
2	Data Code - use 1
4 - 9	Table Name - use CTABLE
25 - 36*	Velocity Increment (ft/sec)
73 - 80	Card No/Identification: Available for User

Routing Coefficient Cards

Column(s)	Content
2	Data Code - use 8
13 - 24*	
25 - 36	
37 - 48	Data Field: Enter C coefficients
49 - 60	
61 - 72	
73 - 80	Card No//Identification

ENDTBL Card

Column(s)	Content
2	Data Code - use 9
4 - 9	Table Name - use ENDTBL
73 - 80	Card No./Identification: Available for User

* NOTE: Punch decimal point but do NOT include comas to indicate thousands

** NOTE: Must be included in every run.

TABLE 29

ROUTING COEFFICIENT TABLE C VS. VELOCITY

SCS-598
S-64

Watershed _____ Hydrologist _____ Date _____

IMPORTANT: Line out unused cards. Data fields require decimal points. KEYPUNCHER: Left justify data fields.

Enter successive entries left to right with initial entry for velocity = 0. Fill last row of data with last entry of table.

VELOCITY INCREMENT, FT/SEC

DATA CODE	TABLE NAME	DATA FIELD #1	DATA FIELD #2	DATA FIELD #3	DATA FIELD #4	DATA FIELD #5	CARD NO./IDENTIFICATION
	CTABLE	0.20					1
		.00	.08	.18	.25	.32	2
		.37	.41	.45	.49	.51	3
		.54	.57	.59	.61	.63	4
		.65	.66	.67	.69	.70	5
		.71	.72	.73	.74	.75	6
		.76	.77	.77	.78	.79	7
		.79	.80	.81	.81	.82	8
		.82	.83	.83	.84	.84	9
		.84	.85	.85	.86	.86	10
		.86	.86	.87	.87	.87	11
		.88	.88	.88	.89	.89	12
		.89	.89	.89	.89	.90	13
		.90	.90	.90	.90	.91	14
		.91	.91	.91	.91	.91	15
		.92	.92	.92	.92	.92	16
		.92	.92	.92	.93	.93	17
	ENDTBL						18

Line/card Velocity "c"
ft/sec
2   0   0
     .2   0.08
     .4   .18
     .6   .25
     .8   .32
3   1.0   .34
     1.2   .41
     1.4   .45
     1.6   .49
     1.8   .51
4   2.0   .54
     2.2   .57
     2.4   .59
     2.6   .61
etc etc   etc

NOTE: This card must be the last card of this table.

WATERSHED PROGRAM, SOIL CONSERVATION SERVICE, SEPT. 28, 1962

table is included in columns 25-36 of the CTABLE card. The C values given by Eq. 33 are given for values of the velocity starting at 0 fps and incremented by the constant given on the CTABLE card. Table 29 shows the Routing Coefficient Table for an increment of 0.2 fps.

The Dimensionless Hydrograph Table

The Dimensionless Hydrograph Table gives the ordinates of the curvilinear unit hydrograph for a constant time increment. Both the discharge and time axes are dimensionless. The discharge axis is given as a ratio of the discharge to the peak discharge; therefore, the peak ordinate equals 1. The time axis is given as a ratio of the time to the length of the time base; this differs from the previous discussion in that the time axis had been given as a ratio of time to the time-to-peak. Because the time-to-peak is one-fifth the time base, the peak discharge of the unit hydrograph occurs at a dimensionless time of 0.2.

Table 30 summarizes the input required for the Dimensionless Hydrograph Table. The tabular data is preceded by a DIMHYD card and followed by a ENDTBL card. The dimensionless time increment must be specified in columns 25-36 of the DIMHYD card. The tabular data are given in constant increments of dimensionless time, with the first ordinate for a time of zero. As an example Table 31 gives the standard SCS Dimensionless Hydrograph Table.

Cumulative Rainfall Tables

There are three types of rainfall tables, which are labeled in TR-20 as follows: 1) the One-day Watershed Evaluation Storm, 2) Emergency Spillway or Freeboard Hydrograph; and 3) Actual. These names reflect their intended use. The three types may be easier understood if they are classified on the basis of the scale used for the precipitation and time axes. Table 32 summarizes the scales of the rainfall tables and indicates the input that is required to reference the tables. It is especially important to emphasize that each table must be expressed in cumulative form.

The One-Day Watershed Evaluation Storm. Table 32 indicates that the precipitation scale for this cumulative rainfall table is dimensionless and the time scale ranges from 0 to 24 hours. This table is, therefore, of use when the standard type I or II is to be used. It is assigned a table number of 1, which is indicated in column 11 of the RAINFL card. Because the precipitation scale is dimensionless, an actual rainfall volume in inches must be specified on the COMPUT card; also, because an actual time scale is used, a duration of 1.0 is specified on the COMPUT card. The actual rainfall volume and the duration of 1.0 serve as multipliers to convert the rainfall table to a dimensional form. This rainfall table is suitable for the evaluation of watersheds in which the travel time through the watershed is approximately two days or less.

Table 33 summarizes the card images that are necessary to create the One-Day Watershed Evaluation Rainfall Table. The rainfall ratio cards contain the dimensionless ordinates of the rainfall scale, and they are given at the time increment specified on the RAINFL card. The rainfall ratio cards are preceded by the RAINFL card and followed by an ENDTBL card. An example of these cards is shown in Table 34; this gives the ordinates of the SCS type I storm.

TABLE 30

DIMENSIONLESS-HYDROGRAPH TABLE[**]

Column(s)	Content
2	Data Code - use 4
4 - 9	Table Name - use DIMHYD
25 - 36[*]	Dimensionless Time Increment (Last entry must be for dimensionless time = 1.0)
73 - 80	Card No. Identification: Available for User

DISCHARGE RATIO CARDS [***]

Column(s)	Content
2	Data Code - use 8
13 - 24[*] 25 - 36 37 - 48 49 - 60 61 - 72	Data Field - Enter discharge ratios
73 - 80	Card No./Identification: Available for User

ENDTBL Card

Column(s)	Content
2	Data Code - use 9
4 - 9	Table Name - use ENDTBL
73 - 80	Card No./Identification: Available for User

[*]   NOTE: Punch decimal point but do NOT include commas to indicate thousands.

[**] NOTE: Must be included in every run   Ref: NEH-4, Chapter 16

[***] NOTE: Number of entries must not exceed 75.

TABLE 31

DIMENSIONLESS HYDROGRAM TABLE, DISCHARGE VS. TIME

Watershed _____ Hydrologist _____ Date _____

IMPORTANT: Line out unused cards. Data fields require decimal points. KEYPUNCHER: Left justify data fields.

DIMENSIONLESS TIME INCREMENT (Last entry must be for dimensionless time 1.0)

Enter successive entries left to right with initial entry for time = 0. Fill last row of data with last entry of table.

DATA CODE	TABLE NAME	DATA FIELD #1	DATA FIELD #2	DATA FIELD #3	DATA FIELD #4	DATA FIELD #5	CARD NO./ IDENTIFICATION
			0.02				19
		.000	.030	.100	.190	.310	20
		.470	.660	.820	.930	.990	21
		1.000	.990	.930	.860	.780	22
		.680	.560	.460	.390	.330	23
		.280	.207	.147	.107	.007	24
		.055	.040	.029	.021	.015	25
		.011	.005	.000			26
							27
							28
							29
							30
							31

Ref. Chapter 16

Time Ratio T/Tp	Discharge Ratio q/qp
0	0
.02	
.04	
.06	
.08	
.10	
.12	
.14	
.16	
.18	
.20	
.22	
.24	
.26	
.28	
.30	
etc	

WATERSHED PROGRAM, SOIL CONSERVATION SERVICE SEPT. 28, 1965

TABLE 32. Summary of Precipitation Tables

Type	Rainfall Table No.	Precipitation Scale	Time Scale	COMPUT Card Input	
				Rainfall Volume	Rainfall Duration
Cumulative Rainfall Table: One-day Storm	1	dimensionless (0 to 1)	actual (24 hours)	actual (inches)	1.0
Cumulative Rainfall Table: Emergency Spillway	2	dimensionless (0 to 1)	dimensionless (0 to 1)	actual (inches)	actual (hours)
Cumulative Rainfall Table: Actual	3-9	actual (inches)	actual (hours)	1.0	1.0

95

TABLE 33

CUMULATIVE RAINFALL TABLE: One-Day Storm

RAINFL Card

Column(s)	Content
2	Data Code - use 5
4 - 9	Table Name - use RAINFL
11	Table No. - use 1
25 - 36*	Time Increment (hours)
73 - 80	Card No./Identification: Available for User

RAINFALL RATIO CARDS

Column(s)	Content
2	Data Code - use 8
13 - 24*	
25 - 36	
37 - 48	Data Field - Enter rainfall ratios
49 - 60	
61 - 72	
73 - 80	Card No./Identification: Available for User

ENDTBL Card

Column(s)	Content
2	Data Code - use 9
4 - 9	Table Name - use ENDTBL
73 - 80	Card No./Identification: Available for User

*Note: Punch decimal point but do NOT include commas to indicate thousands.

TABLE 34

CUMULATIVE RAINFALL TABLE, FOR ONE-DAY SCS TYPE I PRECIPITATION DISTRIBUTION

SCS-272(a)
Rev.

Watershed _____   Hydrologist _____   Date _____

DATA CODE	TABLE ID					CARD NO. / IDENTIFICATION
	NAME	NO.				

IMPORTANT: Line out unused cards. Data fields require decimal points. KEYPUNCHER: Left justify data fields.

	DATA FIELD #1	DATA FIELD #2	DATA FIELD #3	DATA FIELD #4	DATA FIELD #5	CARD NO.
TIME INCREMENT*	0.5					32

TABLE NO. 1

5 RAINFL

Enter successive entries left to right with first entry for time = 0. Fill last row of data with last entry of table.

DATA FIELD #1	DATA FIELD #2	DATA FIELD #3	DATA FIELD #4	DATA FIELD #5	CARD NO.
.000	.008	.017	.026	.035	33
.045	.055	.063	.076	.087	34
.099	.112	.125	.140	.156	35
.174	.194	.219	.254	.303	36
.515	.583	.624	.654	.682	37
.705	.727	.748	.767	.784	38
.800	.816	.830	.844	.857	39
.870	.882	.893	.905	.916	40
.926	.936	.946	.955	.965	41
.974	.983	.992	1.000	1.000	42

NOTE: This card must be the last card of the table.

9 ENDTBL   43

WATERSHED PROGRAM, SOIL CONSERVATION SERVICE, SEPT 30, 1963

*Time increment is 0.5 hour. On "Executive Control for Watershed" (SCS-274) form set DATA FIELD #2 to actual rainfall depth and rainfall duration, DATA FIELD #3, to 1.0.

97

Emergency Spillway or Freeboard Hydrograph Table. Table 32 indicates that both the precipitation and time scales for this table are dimensionless. Therefore, the actual rainfall volume and rainfall duration must be specified on the COMPUT card; these values are used as multipliers to form a dimensional table. This table is assigned a value of 2 for the rainfall table number. Tables 35 and 36 show the input and an example, respectively.

Actual Storm. Table 32 indicates that both axes of this cumulative rainfall table are given in actual amounts, with the precipitation volume in inches and the storm duration in hours. Thus, both scaling factors on the COMPUT card are set at 1.0, which indicates that scaling is unnecessary because the input is already dimensional.

Table 37 summarizes the card images that are necessary to construct this rainfall table. The number of rainfall depth cards can not exceed 20 (i.e., 100 field widths), and all five field widths on each card, including the last, must be filled in. Table 38 shows a Cumulative Rainfall Table: Actual for an actual 24-hour storm having a total depth of 4 inches; the cumulative depths are given in increments of 2 hours, which is indicated on the RAINFL card. The first entry in the rainfall depth cards is for a time of zero hours (i.e., the start of precipitation).

Stream Cross-Section Data Table

A stream cross-section data table is required for each channel reach that is numbered in columns 13-15 of the standard control. The data table summarizes the water surface elevation-discharge-cross sectional end area relationship. It consists of a series of cross-section data cards that are preceded by a XSECTN card and followed by an ENDTBL card. There can not be more than 20 cross-section data cards for any one cross-section. Values between the points that are used to define the cross-section are estimated using straight line interpolation. For extrapolation beyond the highest elevation, a straight line extension is made using the last two values of data shown. If the discharge is given in cubic feet per second per square mile (CSM), then the total drainage area above the cross-section must be shown on the XSECTN card. The computer multiplies the value shown in this space by the discharge in CSM in order to convert to cfs. If the discharge is given in cfs, then a value of 1.0 must be inserted for the drainage area on the XSECTN card (Table 39).

Table 40 shows the card images that comprise the cross-section data table. The drainage area above cross-section 001 is 5.4 square miles, as indicated on the XSECTN card. Because a drainage area is given on the XSECTN card, the discharge values on the cross-section data cards are given in CSM.

Preparation of XSECTN Tabular Data. The preparation of the XSECTN tabular data can be a very important step in a TR-20 modeling effort. Frequently, the objective of the TR-20 modeling is to determine flood plain elevations; the flood plain elevations are computed on the basis of the XSECTN table data. Also, the accuracy of stream reach routing is dependent upon the accuracy of the XSECTN tables. Therefore, the success of a TR-20 modeling effort may be highly dependent on the accuracy of the XSECTN tabular data.

If the cross-sections in a particular stream reach are fairly uniform

98

TABLE 35

CUMULATIVE RAINFALL TABLE, for Emergency Spillway or Freeboard Hydrograph Storms

RAINFL Card

Column(s)	Content
2	Data Code - use 5
4-9	Table Name - use RAINFL
11	Table No. - use 2
25-36*	Time Increment (hours) for rainfall ratio cards
73-80	Card No/Identification: Available for User

RAINFALL RATIO Cards   (same as for One-Day Storm Table)

ENDTBL Card   (Same as for the One-Day Storm Table)

---

*Note: Punch decimal point but do NOT include commas to indicate thousands.

TABLE 36

CUMULATIVE RAINFALL TABLE, FOR EMERGENCY SPILLWAY OR FREEBOARD HYDROGRAPHS

SCS-272(b)
Rev.

Watershed _____ Hydrologist _____ Date _____

DATA CODE · TABLE ID (NAME · NO.)

IMPORTANT: Line out unused cards. Data fields require decimal points. KEYPUNCHER: Left justify data fields.

TIME INCREMENT* : .02

Enter successive entries left to right with first entry for time = 0. Fill last row of data with last entry of table.

Card No./Identification	Data Field #1	Data Field #2	Data Field #3	Data Field #4	Data Field #5
45	.00	.01	.02	.02	.03
46	.04	.05	.06	.07	.08
47	.10	.11	.13	.14	.17
48	.19	.22	.27	.35	.44
49	.52	.60	.63	.66	.68
50	.70	.72	.74	.76	.77
51	.79	.80	.82	.83	.84
52	.85	.87	.88	.89	.90
53	.91	.92	.93	.94	.95
54	.9567	.9633	.97	.98	.99
55	1.00	1.00	1.00	1.00	1.00

NOTE: This card must be the last card of the data table.

WATERSHED PROGRAM, SOIL CONSERVATION SERVICE, SEPT. 31, 1965

*Time increment is 0.02 of unit duration, hence storm duration and rainfall depth need to be shown in DATA FIELDS #3 and #2 respectively on "Executive Control for Watershed" (SCS-274) form.

100

TABLE 37

CUMULATIVE RAINFALL TABLE for Actual Storms

RAINFL Card

Column(s)	Content
2	Data Code - use 5
4-9	Table Name - use RAINFL
11	Table No. - use an integer from 3 to 9
25-36*	Time increment (hours) for rainfall depth cards
73-80	Card No/Identification: Available for User

RAINFALL DEPTH Cards

Column(s)	Content
2	Data Code - use 8
13-24* 25-36 37-48 49-60 61-72	Rainfall depths (in inches)
73-80	Card No/Identification - Available for User

ENDTBL Card (Same as for the One-Day Storm Table)

---

*Note: Punch decimal point but do NOT include commas to indicate thousands.

TABLE 38

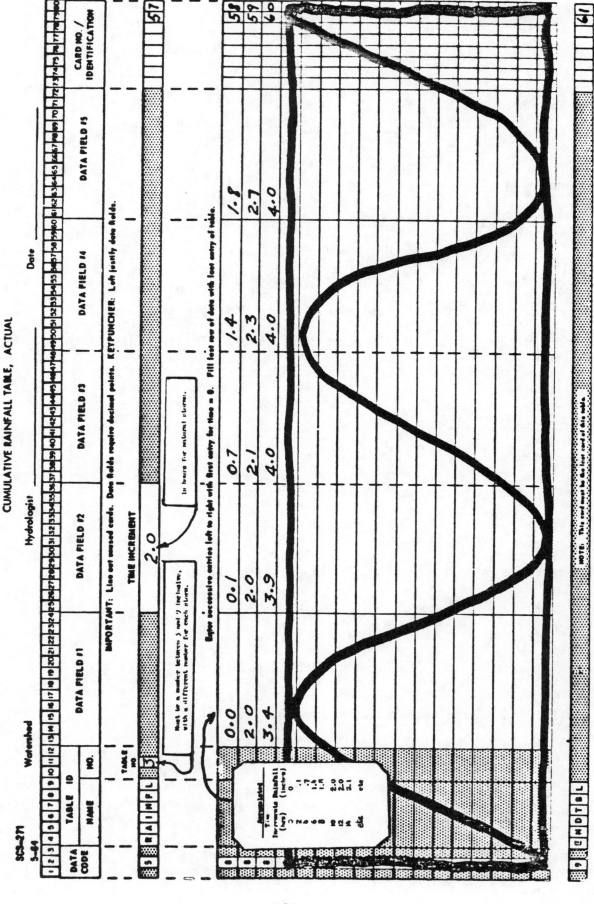

TABLE 39

STREAM CROSS-SECTION DATA TABLE

XSECTN Card

Column(s)	Content
2	Data Code - use 2
4 - 9	Table Name - use XSECTN
13 - 15	Cross-section number (001 to 120)
25 - 36*	Drainage Area (square miles)**
73 - 80	Card No./Identification: Available for user.

CROSS-SECTION Data Cards

Column(s)	Content
2	Data Code - use 8
25 - 36*	Elevation (ft)
37 - 48*	Discharge (csm or cfs) **
49 - 60*	End Area (square feet)
73 - 80	Card No./Identification: Available for user.

ENDTBL Card

Column(s)	Content
2	Data Code - use 9
4 - 9	Table Name - use ENDTBL
73 - 80	Card No./Identification: Available for User

* NOTE: Punch decimal point but do NOT include commas to indicate thousands.

** NOTE: If discharge is given in cubic feet per second per sq. mile (csm) the total drainage area above the cross section must be shown on the XSECTN card. The computer multiplies the value shown in this space by the discharge in csm in order to convert to csf. If the discharge is given in cfs, a value of 1.0 must be inserted for the drainage area.

TABLE 40

STREAM CROSS-SECTION DATA   Cross Section No. _____

Watershed _____   Hydrologist _____   Date _____

SCS-270
5-64

IMPORTANT: Line out erased cards. Data fields require decimal points. KEYPUNCHER: Left justify data fields.

DATA CODE	TABLE NAME	DATA FIELD #1	DATA FIELD #2	DATA FIELD #3	CARD NO / IDENTIFICATION

X SECTN NO (001 - 701)

DRAINAGE AREA, SQ MI   5.4

(If rebalanced discharge in CFS, enter 1 0)

ELEVATION, FT.	DISCHARGE, CSM	END AREA, SQ. FT	Card No.
742.	0.0	0.0	63
743.	8.0	70.	64
744.	20.0	80.	65
746.	75.	190.	66
748.	200.	350.	67
750.	450.	650.	68
752.	800.	1350.	69
754.	1400.	2450.	70
755.	1800.	3150.	71

62

72

NOTE: This card must be the last card for each cross section.

WATERSHED PROGRAM SOIL CONSERVATION SERVICE JAN 30 1964

104

throughout and the slope is fairly constant, then Manning's formula may be used for computing the XSECTN table. A typical cross-section in the reach can be selected and the steady state discharges, elevations, and end areas can be computed directly from Manning's formula. For instance, several elevations can be selected, the end areas can be computed on the basis of the channel geometry, and the steady-state discharge can be computed from Manning's formula.

If the cross-sections or slopes in a particular stream reach are not uniform or there are obstructions, such as bridges or culverts, then the determination of the XSECTN tabular data can involve an extensive level of effort. In these cases, a backwater computation program, such as the SCS's WSP2 or the Corps of Engineers HEC-2, should be used. The preparation of the input data deck for the backwater curve program may require that a field survey be performed. Typical input data requirements for backwater curve programs include:

1. stream cross-section geometry at several points in the stream reach,

2. bridge and culvert geometry data,

3. Manning's "n" values along the entire length of the reach, and

4. elevation-discharge data at the appropriate control point.

The backwater program should be run using several different discharges. The XSECTN table can then be determined by selecting a typical cross-section in the reach and using the resultant end areas and elevations at each of the discharges used. The selection of the "typical" cross-section is very important and, therefore, should be done with great care.

Structure Data Table

A Structure Data Table is required for each structure number identified in columns 16-17 of the standard control. The table summarizes the water surface elevation-discharge-reservoir storage relationship. It consists of a series of elevation-discharge-storage cards preceded by a STRUCT card and followed by an ENDTBL card. The number of coordinates that are used to describe a structure can not exceed 20 elevations. The zero discharge on the first card image must be oriented to the crest elevation of the low stage outlet in the principal spillway (i.e., the elevation on the RESVOR card of standard control at time zero). A maximum of 99 structures can be used. The structure number placed in columns 16-17 of the STRUCT card will correspond to the structure number located in columns 16-17 of the RESVOR card in standard control.

Table 41 summarizes the card images that are used to form a Structure Data Table. Table 42 provides an illustration.

Null Structure. Many problems involve an analysis in which one objective is to determine the effect of a structure. In such cases, a Structure Data Table that includes just one elevation-discharge-storage card can be created; this card includes the elevation of zero discharge, a value of zero for the discharge, and the corresponding storage. The problem would be analyzed by executing the standard control using the null structure table and then executing the Modify Standard Control with the RESVOR card altered

TABLE 41

STRUCTURE DATA TABLE

STRUCT Card

Column(s)	Content
2	Data Code - use 3
4 - 9	Table Name - use STRUCT
16 - 17	Structure No. (01 to 60)
73 - 80	Card No./Identification: Available to user.

Elevation - Discharge - Storage Cards **

Column(s)	Content
2	Data Code - use 8
25 - 36*	Elevation (feet) ***
37 - 48*	Discharge (cfs)
49 - 60*	Storage (acre-feet)
73 - 80	Card No./Identification: Available to user.

ENDTBL Card

Column(s)	Content
2	Data Code - use 9
4 - 9	Table Name - use ENDTBL
73 - 80	Card No./Identification: Available to User

* NOTE: Punch decimal point but do NOT include commas to indicate thousands.

** NOTE: The number of coordinates that are used to describe a structure cannot exceed the data field spaces on a single input form (20 elevations).

***NOTE: The zero discharge on the first line must be oriented to the crest elevation of the low stage outlet in the principal spillway (Surface Elevation at T = 0).

106

TABLE 42

STRUCTURE DATA

SCS-239
5-64

Watershed _____ Hydrologist _____ Date _____

Structure No. _____

DATA CODE	TABLE NAME	DATA FIELD #1	DATA FIELD #2	DATA FIELD #3	CARD NO / IDENTIFICATION
	STRUCT				180

STRUCTURE NO (01.60) 05

IMPORTANT: Line out unused cards. Data fields require decimal points. KEYPUNCHER: Left justify data fields.

ELEVATION, FT	DISCHARGE, CFS	STORAGE, ACRE FT	CARD NO / IDENTIFICATION
663.	00	200.	181
664.	58.	250.	182
668.	256.	375.	183
672.	300.	575.	184
676.	352.	860.	185
680.	371.	1225.	186
684.	396.	1650.	187
688.	418.	2200.	188
690.4	440.	2575.	189
691.4	1286.	2740.	190
692.4	3440.	2900.	191
693.4	6802.	3075.	192
694.4	10950.	3250.	193
695.4	15677.	3425.	194
696.4	21034.	3600.	195

END TBL   196

NOTE: This card must be the last card for each structure

WATERSHED PROGRAM SOIL CONSERVATION SERVICE JAN 21 1964

107

so that the elevation-discharge-storage relationship for the proposed structure is used.

Discharge Hydrograph Table

The Discharge Hydrograph Table provides the means of inputting a hydrograph. Thus, it serves as an alternative to using the RUNOFF subroutine to develop a hydrograph. The Discharge Hydrograph Table consists of two READHD cards followed by a series of Discharge Cards and an ENDTBL card. The first READHD card indicates the machine storage element number (1 to 7) into which the hydrograph will be placed; this READHD card has an 8 in column 11 while the second READHD card has a 9 in column 11. The content of the Discharge Hydrograph Table is given in Table 43 and an example is shown in Table 44. The hydrograph is limited to 300 ordinates.

EXECUTIVE CONTROL

The Executive Control serves both to execute the standard control and to provide data that are necessary for execution. Executive control consists of three subroutine operations: the INCREM card, the COMPUT card, and the ENDCMP card. The Executive Control cards are placed after both the tabular data and the Standard Control.

INCREM Card

The sole purpose of the INCREM card is to specify the main time increment in hours (columns 25-36). All hydrographs generated by the program will be calculated using this time increment. It is important that the main time increment be made short enough to adequately describe the hydrographs for smaller drainage areas and large enough that, when multiplied by the number of coordinates, it will extend through the peak periods. An INCREM card must precede the first COMPUT card, and it remains in force until superceded by the insertion of a new INCREM card. The content of the INCREM card is given in Table 45.

COMPUT Card

The COMPUT card specifies: 1) the cross-sections or structure locations where routings are to begin and end; 2) the rainfall starting time, depth, and duration; 3) the rain table number; and 4) the antecedent soil moisture condition (I, II, or III). The contents of a COMPUT card are summarized in Table 46. The rainfall depth and duration are placed in columns 37-48 and 49-60, respectively. If actual rainfall depths and times are given on the Cumulative Rainfall Table for either the depth or the duration, then a 1.0 is inserted into appropriate columns. If either the rainfall depth or duration is dimensionless on the Cumulative Rainfall Table, then the actual depth or duration is given on the COMPUT card. The rain table number is placed in column 65 of the COMPUT card.

ENDCMP and ENDJOB cards

After the Executive Control cards are completed, an ENDCMP card should follow. The content of the ENDCMP card is given in Table 47. If the Standard Control is modified, then appropriate Executive Control cards must follow. After all computations are complete, the program deck must be ended

TABLE 43

READ-DISCHARGE-HYDROGRAPH DATA TABLE

READHD Card No. 1

Column(s)	Content
2	Data Code - use 7
4 - 9	Table Name - use READHD
11	Card Indicator - use 8
17	Computer Storage Element No. (1 to 7)
73 - 80	Card No./Identification: Available to user.

READHD Card No. 2

Column(s)	Content
2	Data Code - use 7
4 - 9	Table Name - use READHD
11	Card Indicator - Use 9
13 - 24*	Starting Time (hours)
25 - 36*	Time Increment (hours)
37 - 48*	Drainage Area (square miles)
49 - 60*	Base flow (cfs)
73 - 80	Card No./Identification: Available for user.

Discharge Cards

Column(s)	Content
2	Data Code - use 8
13 - 24* 25 - 36* 37 - 48* 49 - 60* 61 - 72*	Discharge (cfs), with first value in table relating to starting time shown on READHD Card No. 2
73 - 80	Card No./Identification/Available to user.

* NOTE: Punch decimal point but do NOT include commas to indicate thousands.

TABLE 44
READ DISCHARGE HYDROGRAPH

SCS-276
5-64

Watershed _____ Hydrologist _____ Date _____

DATA CODES	TABLE NAME	DATA FIELD #1	DATA FIELD #2	DATA FIELD #3	DATA FIELD #4	DATA FIELD #5	CARD NO./ IDENTIFICATION
7	READ MD 8						276
	LOCATION	6					
7	READ MD 9	STARTING TIME, HRS. 0.0	TIME INCREMENT, HRS. 2.0	DRAINAGE AREA, SQ. MI. 26.84	BASE FLOW, CFS		277
		0.0	100.	300.	550.	1350.	278
		1900.	1800.	1200.	950.	700.	279
		500.	300.	225.	250.	700.	280
		1450.	1350.	1100.	925.	550.	281
		625.	575.	525.	500.	600.	282
		1000.	775.	600.	400.	400.	283
		750.	500.	325.	300.	300.	284
		300.	300.	275.	225.	175.	285
		125.	90.	80.	50.	40.	286
		30.	25.	20.	15.	10.	287
		5.	0.0	0.0	0.0	0.0	288
9	END TBL						289

IMPORTANT: Line out unused cards. Data fields require decimal points. KEYPUNCHER: Left justify data fields.

Places the hydrograph into computer storage element 6.

Enter successive entries left to right with first entry for starting time. Fill last row of data with last entry of table.

Zero discharge is related to zero starting time. Starting time on line/card 277 will relate to starting time on line/card 291.

Discharge rate is described from left to right in time increments of 2 hours.

This gaged hydrograph is inserted ahead of line/card 590 as the inflow hydrograph to str. 06.

NOTE: This card must be the last card of this table.

WATERSHED PROGRAM, SOIL CONSERVATION SERVICE, SEPT. 30, 1963

110

TABLE 45

Column(s)	Content
2	Data Code - Use 7
4-9	Indicator - Use INCREM
11	Operation Number - Use 6
25-36*	Time Increment (hours)
73-80	Card No. Identification: Available for User

* Note: punch decimal point but do NOT include commas to indicate thousands

TABLE 46

COMPUT   Card **

Column(s)	Content
2	Data Code - Use 7
4-9	Indicator - use COMPUT
11	Operation Number - Use 7
13-17	Cross-section number (13-15) or structure number (16-17) where computation begins
19-23	Cross-section number (19-21) or structure number (22-23) where computation ends (Note: computation includes this X-section or structure)
25-36*	Time at which computation starts (hours)
37-48*	Rainfall depth (inches)
49-60*	Rainfall duration
65	Rain Table No. (1 to 9)
69	Soil (1:Dry; 2:Normal; 3:Wet)
73-80	Card No. Identification: Available for User

* Note: punch decimal point but do NOT include commas to indicate thousands

**Note: more than one COMPUT card can be used when different precipitation or soil moisture conditions exist within a watershed.

TABLE 47.   The ENDCMP Card

Column(s)	Content
4-9	Indicator: Use ENDCMP
11	Operation Number - Use 1
73-80	Card No. Identification: Available for user

TABLE 48.   The ENDJOB Card

Column(s)	Content
4-9	Operation - insert ENDJOB
11	Operation No. - insert 2
73-80	Card No. Identification: Available for user

TABLE 49.   The INSERT Card

Column(s)	Content
2	Data Code - use 7
4-9	Operation Name - use INSERT
11	Operation No. - use 2
13-15*	Cross-section numbers: the cards following the INSERT card are inserted in Standard Control following the first card that uses this X-Section Number
16-17*	Structure Number: the cards following the INSERT Card are inserted in Standard Control following the first card that uses the Structure no.
73-80	Card No. Identification: Available for user

*Note: Insert either a cross-section no. or a structure no., but not both.

with an ENDJOB card, the contents of which are specified in Table 48.

MODIFY STANDARD CONTROL

The Modify Standard Control subroutines are used to insert new routines in the Standard Control sequence, to alter the input data for existing operations, and to delete any operations that are already in the sequence. Many hydrologic analyses are intended to examine various alternatives, such as changes in land use or the insertion of a control structure into the watershed system. Problems are solved most effectively on a computer when all alternatives are examined in one run rather than one computer run per alternative. The Modify Standard Control operations enable the user to modify the original Standard Control sequence and analyze an alternative. A separate Executive Control sequence is placed after each Modify Standard Control sequence.

There are six Modify Standard Control subroutine operations:

1. INSERT: an operation used to insert new Standard Control subroutines into the Standard Control sequence;

2. ALTER: an operation that is used to change any input on a Standard Control card image, including machine storage numbers, data field input, and output options;

3. DELETE: an operation that causes one or more Standard Control operations to be deleted from the Standard Control sequence;

4. LIST: an operation that causes the Tabular and Standard Control data to be printed out;

5. UPDATE: an operation that is used when the library tape for a watershed is to be retained for subsequent processing, and it is desired that all new Tabular Data and modifications to Standard Control operations should replace the original counterparts to become permanent record; and

6. BASFLO: an operation that allows a uniform rate of base flow (cfs) to be introduced into reach routings at any location.

The INSERT Card

TR-20 is quite frequently used to evaluate the effect of structures in a developing watershed. This use represents one example where an INSERT card would be of value. A standard sequence of Tabular Data, Standard Control, and Executive Control would be used to analyze the watershed without the structure. Then the Modify Standard Control and necessary Executive Control card images could be used to analyze the watershed with the structure.

The contents of an INSERT card are summarized in Table 49. The INSERT card is followed by one or more Standard Control cards. These cards are inserted into the Standard Control sequence immediately following the first

114

reference to a cross-section number or a structure-number having the same value as that which is located in columns 13-15 or 16-17, respectively, of the INSERT card.

The ALTER Card

The ALTER card indicates that one or more card images in the Standard Control sequence are to be altered. Following the ALTER card is a sequence of Standard Control card images. These cards will replace the card images in the Standard Control sequence that are identical to the values in columns 1 to 18. The input for the ALTER card is shown in Table 50.

The DELETE Card

The DELETE card is used to eliminate card images from the Standard Control sequence. The DELETE card is followed by one or more Standard Control card images; the cards in the Standard Control sequence that are identical in columns 1-18 to these cards are eliminated. Table 51 provides a summary of the contents of the DELETE card.

The BASFLO Card

The BASFLO card is used to add a constant baseflow (cfs) to each hydrograph. If a second BASFLO card is inserted, the difference between the old baseflow and the new baseflow is added to the inflow hydrograph when the next REACH subroutine is referenced. Table 52 summarizes the input for the BASFLO card.

EXECUTION SEQUENCE

In general, a TR-20 program will be structured with the Tabular Data first and then the Standard Control and Executive Control operations. If additional analyses are to be executed using Modify Standard Control operations, then the modifications follow the ENDCMP card. An Executive Control sequence will then follow the Modify Standard Control operations. When additional sequences of Modify Standard Control-Executive Control are not included, then the program run is terminated with an ENDJOB card. A comprehensive Example is provided in Appendix A.

CONTROL CARDS

In order to use TR-20 it will be necessary to access the program using control cards. It is, first, necessary to distinguish between two types of control cards, internal and external. The external control cards will not be discussed herein because they are machine dependent; however, a description of the necessary cards are easily obtained from the computer programmer who placed TR-20 on the computer system.

The internal control cards are the JOB and TITLE cards. They are placed at the very start of the TR-20 deck; examples of the cards are given in Appendix A.

For the JOB card, the words "JOB" and "TR-20" are entered in columns 1-3 and 6-9, respectively; this is all the input that is required. However, additional out-put can be obtained by entering "FULLPRINT" in columns 31-39;

selection of this option will result in a more detailed output. Other options are available and are outlined in the TR-20 manual; they will not be discussed herein because they are rarely used.

The TITLE card, which follows the JOB card, allows the user to include a title on the output; the title is often a short identifier so that the output can be readily identified. The word "TITLE" is placed in columns 1-5. An optional run number can be entered in columns 7-9. Columns 11-80 are available to the user for the title; they are optional and can be left blank. However, if a title is not desired, the user must still include the TITLE card.

TABLE 50.   The ALTER Card*

Column(s)	Content
2	Data Code - use 7
4-8	Operation Name - use ALTER
11	Operation No. - use 3
73-80	Card No. Identification: Available for user

*Note: Data given in columns 1-18 on cards following the ALTER card must be identical with Standard Control card to be altered.

TABLE 51.   The DELETE Card*

Column(s)	Content
2	Data Code - use 7
4-9	Operation Name - use DELETE
11	Operation No. - use 4
73-80	Card No. Identification: Available for user

*Note: Data given in columns 1-18 of the cards following the DELETE catd must be identical with Standard Control card to be deleted.

TABLE 52.   The BASFLO Card*

Column(s)	Content
2	Data Code - use 7
4-9	Operation Name - use BASFLO
11	Operation No. - use 5
25-36	New Base Flow (cfs)
73-80	Card No. Identification: Available for user

*Note: This card usually precedes a COMPUT Card.

SECTION 17

STORMWATER MANAGEMENT

Stormwater Management Basin Design with TR-55

It is recognized that the urbanization process is responsible for increases in runoff volumes and peak discharge rates, as well as sediment loads. The concept of stormwater management (SWM) is best understood through a superficial examination of the urbanization process. Urban development increases the percent of the drainage area that is covered by impervious surfaces and changes the drainage pattern such that the flow lengths are decreased. These physical changes (i.e., reductions in interception, depression storage, and infiltration potential) have the effect of decreasing the natural storage of the catchment. The intent of SWM is to mitigate the hydrologic impacts of this lost natural storage using man-made storage. The volume of man-made storage that is allocated to the SWM basin will not necessarily equal the volume of natural storage that is lost because both the urbanization process and the SWM basin change the time characteristics of runoff, as well as the volume.

To mitigate these detrimental effects, SWM policies have been adopted with the intent of limiting peak flow rates from developed areas to that which occurred prior to development. While other SWM methods, both structural and nonstructural, are available, SWM basins (i.e., detention/retention basins) appear to be the most popular.

Chapter 7 of TR-55 provides two methods for estimating the volume of detention storage that is necessary to control the peak flow rate after development to that peak which occurred prior to development. The two methods are not alternatives, but instead, they are used for two different design conditions. The criteria used to select one of the two procedures is the allowable flow rate. The inflow-outflow-storage method is recommended when the peak flow rate out of the SWM basin is less than 150 csm (cfs per square mile) for weir flow and 300 csm for pipe flow. When the flow rate exceeds these values, the volume-rate method should be used.

The Inflow-Outflow-Storage Method. As indicated previously, this method is used for the relatively small peak flow rates. The input requirements and computational procedure are specified on the computation sheet of Table 53. After the volume of storage is determined, it is necessary to proportion the outlet facility so that the design outflow rate and the maximum storage occur at the same stage.

TR-55 indicates that estimates made with the inflow-outflow-storage method are within 5 percent for release rates less than 100 csm and within 10 percent for larger release rates.

The Volume-Rate Method. The volume-rate method is used when peak flow rates exceed 150 csm for weir flow or 300 csm for pipe flow. The computational procedure is outlined in Table 53. Fig. 29, which is used to estimate

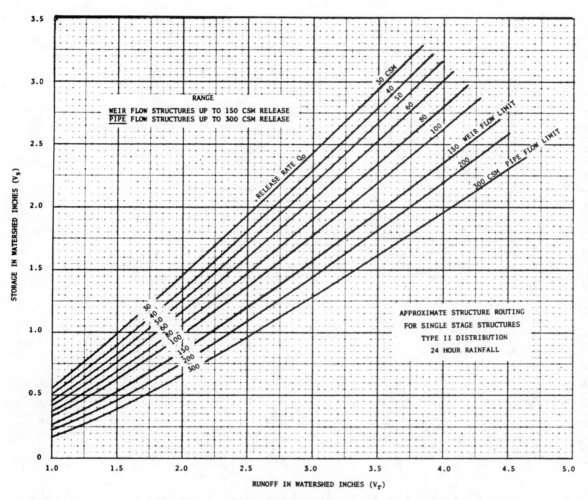

Figure 28. Approximate single-stage structure routing for weir flow structures up to 150 csm release rate and pipe flow structures up to 300 csm release rate.

TABLE 53

COMPUTATION SHEET: VOLUME OF DETENTION STORAGE

1. Input: A = _____ sq. miles : drainage area

   CN = _____ : runoff curve number after development

   P = _____ inches : precipitation depth for design frequency

   Q_o = _____ CSM : release rate from the detention structure

   Structure type (indicate one): pipe flow _____ weir flow _____

2. Select method: _____ Inflow-Outflow-Storage Method: $Q_o \leq$ <150 csm for weir flow or

   $Q_o \leq$ <300 csm for pipe flow

   _____ Volume Rate Method: Q_o >150 csm for weir flow

   Q_o >300 csm for pipe flow

3. Inflow-Outflow-Storage Method:

   a. V_r = _____ inches : determine volume of runoff from Eq. 7 or Fig. 5

   b. V_s = _____ inches : determine volume of storage in watershed inches from Fig. 28

   c. V_s' = _____ acre-feet : V_s' = 53.33 V_s A

4. Volume-Rate Method:

   a. Q_i = _____ csm : peak discharge rate into the detention basin

   b. α = _____ : $\alpha = Q_o/Q_i$

   c. $\beta = V_s/V_r$ = _____ : determine from Fig. 29 using α

   d. V_r = _____ inches : determine volume of runoff from Eq. 7 or Fig. 5

   e. V_s = _____ inches : $V_s = \beta V_s A$

   f. V_s' = _____ acre-feet: V_s' = 53.33 V_s A

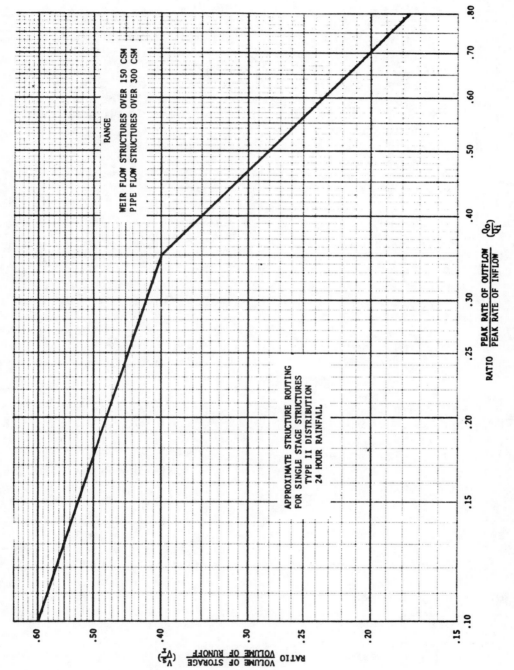

Figure 29. Approximate single-stage structure routing for weir flow structures over 150 csm release rate and pipe flow structures over 300 csm release rate.

the ratio of the volume of storage to the volume of runoff, assumes the volume ratio is between 0.1 and 0.8.

TR-55 indicates that volume ratios (i.e., V_s/V_r) will be up to 25 percent too high when the CN is less than 65 and the time-of-concentration is relatively short. For CN's greater than 85 and long t_c values, the volume ratios from Fig. 29 will be as much as 25% too low.

Example 17-1. A 100-acre watershed, which had a peak discharge of 150 cfs prior to development, is developed. The after development CN is 81. For a precipitation depth of 6 inches, the volume of runoff equals 3.9 inches. The computations are summarized on the computation sheet of Table 54. Because the allowable peak discharge rate exceeds 300 csm the volume-rate method should be used. For a ratio of the before development peak discharge to the after development peak discharge of 0.469, the volume ratio equals 0.30. For 3.9 inches of runoff the volume of storage is 1.17 inches, or 9.73 acre-feet.

Stormwater Management Basin Design with TR-20

The design of SWM basins with TR-20 is generally not a straightforward procedure. The interactions between the inflow hydrographs, basin storage characteristics, outlet structures, and outflow hydrographs can be quite involved. The determination of the optimum SWM basin configuration and the sizes of the outlet structures generally involves an iterative trial and error procedure in which various combinations of basin configuration and outlet sizes are tested via TR-20.

The objective of this section is not to present a comprehensive SWM basin design procedure; the objective is to present some general guidelines and suggestions for determining an optimum SWM basin design.

A logical procedure is to first determine the approximate volume of required storage to meet the SWM objectives. Recent research by the author has indicated that the required volume of storage is not highly dependent on the basin configuration or the outlet structure sizes. Therefore, the approximate volume of storage can be determined independently of the final basin configuration and outlet sizes.

One technique for determining the approximate volume of required storage is to select one outlet size and then develop the elevation-storage data on the basis of rectangular basin configurations; the elevation-discharge data is computed based on the hydraulic characteristics of the outlet structure size. Various sizes of rectangular basins can be tested via TR-20 until a basin size that comes very close to meeting the SWM objective is found. This final basin size is the approximate volume of required storage.

After the approximate storage volume is determined, the basin configuration can be set. The actual basin configuration is a function of the site topography, land costs, soil properties, etc.

The optimum outlet sizes should be determined on a trial basis. Several different outlet sizes should be tested using the actual basin configuration so that the optimum sizes can be found.

TABLE 54

COMPUTATION SHEET: VOLUME OF DETENTION STORAGE

1. Input: $A = \dfrac{100}{640} = 0.156$ sq. miles : drainage area

   $CN = $ __81__ : runoff curve number after development

   $P = $ __6__ inches : precipitation depth for design frequency

   $Q_o = \dfrac{150}{0.156} = $ __962__ CSM : release rate from the detention structure

   Structure type (indicate one): pipe flow _____ ✗ _____ weir flow _____

2. Select method: _____ Inflow-Outflow-Storage Method: $Q_o \leq 150$ csm for weir flow or

   $Q_o \leq 300$ csm for pipe flow

   __✗__ Volume Rate Method: $Q_o > 150$ csm for weir flow

   $Q_o > 300$ csm for pipe flow

3. Inflow-Outflow-Storage Method:

   a. $V_r = $ _____ inches : determine volume of runoff from Eq. 7 or Fig. 5

   b. $V_s = $ _____ inches : determine volume of storage in watershed inches from Fig. 28

   c. $V_s' = $ _____ acre-feet: $V_s' = 53.33\ V_s\ A$

4. Volume-Rate Method:

   a. $Q_i = $ __320__ csm : peak discharge rate into the detention basin

   b. $\alpha = \dfrac{159}{320} = $ __0.449__ : $\alpha = Q_o/Q_i$

   c. $\beta = V_s/V_r = $ __0.30__ : determine from Fig. 29 using α

   d. $V_r = $ __3.9__ inches : determine volume of runoff from Eq. 7 or Fig. 5

   e. $V_s = $ __1.17__ inches : $V_s = \beta V_s A$

   f. $V_s' = $ __9.73__ acre-feet: $V_s' = 53.33\ V_s A$

123

In many jurisdictions, SWM basins are required to control floods of two or more different return periods. In these cases, the outlet structure may contain more than one opening. For instance, an orifice can be used for control of the 2-year flood and a weir at a higher elevation can be used for control of the 10-year flood.

The procedures for SWM basin design for control of multiple return period floods is more involved than for control of only one flood, but the principles are the same. The design procedure discussed above can be employed by designing for each flood successively, starting with the smallest (i.e., shorter return period) flood. For instance the basin configuration and outlet size for the smallest flood can be set first, and the peak elevation attained by this flood can be determined via TR-20. The basin configuration and outlet size above this elevation can be set for control of the next larger flood without affecting the control of the smallest flood.

It should be emphasized that this section presents only some general guidelines and suggestions for design of SWM basins. Because SWM policies are a fairly recent development and each SWM design situation is unique, a comprehensive design procedure does not exist at this time that can be used to meet all SWM policy statements.

APPENDIX A

EXAMPLE: TR-20

Hydrologists and engineers are quite frequently involved in design projects requiring an evaluation of the runoff characteristics before and after development; this is a common analysis required in areas undergoing urban/suburban development. In many locations it is necessary to install an on-site detention reservoir to mitigate the effects of development. Additionally, those involved in development are often required to evaluate the effects of both development and detention storage on downstream properties. Because these analyses are required frequently, the following problem illustrates the use of TR-20 in analyzing both the effect of development and downstream impacts.

Figure A-1 shows a watershed that includes four subareas; the characteristics of each subarea are given in Table A-1 for both the before development and after development conditions. Subarea 1 undergoes development and a detention basin is placed at the outlet from subarea 1. The effect of development and detention is measured on the downstream channel reaches through subareas 2 and 4.

TR-20 Input Sequence

Figure A-2 shows a flowchart of the input sequence for the problem. The program is first run for the before development watershed conditions. After the job control cards and tabular data, the Standard Control sequence is used to define the watershed. The Executive Control sequence, which follows the Standard Control, includes a COMPUT card for both the 100-year and 10-year rainfall depths, both of which follow the INCREM card. A Modify Standard Control sequence is used to indicate the change in land use and the insertion of the structure. A second Executive Control Sequence follows the Modify Standard Control. The ENDJOB card indicates the completion of the job.

Figure A-3 shows a listing of the input deck for the problem. The two job cards, JOB and TITLE, are shown. The tabular data includes the C Table, the dimensionless hydrograph, the rainfall table, the cross section data, and the structure table. Each table is preceded by a definition card (CTABLE, DIMHYD, RAINFL, XSECTN, or STRUCT card) and followed by an ENDTBL card. The C Table and dimensionless hydrograph are not required with many TR-20 programs, as they are internal to the program; they are shown with this example only for illustration of their general structure.

Following the last ENDTBL card, the standard control sequence is shown in Fig. A-3. RUNOFF cards are used for computing runoff hydrographs from subareas 1 and 2. The hydrograph for subarea 1 is then routed through the first channel reach using the REACH card. The reach routed hydrograph and the hydrograph from subarea 2 are then summed using the ADDHYD card. It is especially important to note the machine storage numbers for each operation. The input hydrograph from subarea 1 is stored in unit 5 and is then stored in unit 7 after it is reach routed. The hydrographs in storage units 6 and 7 are then summed and the resulting hydrograph is stored in unit 5. The runoff hydrograph from subarea 3 is determined using the RUNOFF card. It is then added to the hydrograph for subareas 1 and 2. The resulting hydrograph is then reach routed using the REACH card, which is executed after the runoff hydrograph from subarea 4 is computed using a RUNOFF card. The ADDHYD card is then used to obtain the total hydrograph at the outlet. An ENDATA card is used to indicate the end of the Standard Control sequence.

TABLE A-1. Characteristics of Watershed

Subarea	Area (sq mi)	CN	t_c (hrs)
1	0.1563	66	0.70
2	0.1250	71	0.60
3	0.1719	63	0.90
4	0.1172	81	0.40
1 (after development)	0.1563	75	0.55

Channel Reach	Length (ft)
1	1600
2	1150

The Executive Control includes five cards. The first card is used to indicate the computation interval (0.1 hours). The first COMPUT card indicates that the Standard Control sequence from cross section 001 to 005 will be executed using a rainfall depth of 7.0 inches, rainfall table No.1, and antecedent moisture condition 2. The 1.0 in data field 3 indicates that the rainfall table is measured on a real time basis as indicated on the RAINFL card. The second COMPUT card indicates a rainfall depth of 4.3 inches, which is for the 10-yr event. Both COMPUT cards are followed by an ENDCMP card; the first ENDCMP card is necessary in order for the two COMPUT cards to be executed independently of each other.

The Modify Standard Control sequence consists of five cards. The ALTER card is followed by a RUNOFF card which indicates that the CN and time of concentration should be changed to reflect the effect of development. The INSERT card indicates that a structure should be inserted into the execution sequence. The machine storage numbers indicate that the hydrograph stored in unit 5 is routed through the structure; the outflow hydrograph is stored in unit 7. Since the Standard Control sequence indicates that the hydrograph from subarea 1 is stored in unit 5, then it is necessary to use a SAVMOV card to move the structure outflow hydrograph that is stored in unit 7 back to unit 5 before it can be reach routed. A second Executive Control sequence is shown; the INCREM card is not necessary as long as the routing interval is to remain the same. The program run is terminated with the ENDJOB card.

TR-20 Output Sequence

The output for the example of Fig. A-1 is shown in Fig. A-4. The tabular data is printed, along with headings to indicate the contents of the columns. Since a LIST card was included after the ENDATA card, the computer prints the Standard Control cards. The remainder of the output is dependent on the input in columns 61, 63, 65, 67, 69, and 71 on the Standard Control cards. In this case, only peak discharges were requested for all computations except for subarea 1, for which the entire hydrograph was requested. The output includes a description of various input parameters such as the drainage area for the RUNOFF card, the machine storage units for the ADDHYD card, and the reach length for the REACH card; the specific output for each operation is evident from the output for the example.

The Modify Standard Control statements cause a description of the changes to be printed out. The Executive Control then causes the Standard Control, as modified, to be re-run. Again, the output is largely controlled by the output options that were selected by the user.

Summary of Problem Results

This example problem was designed to illustrate the use of TR-20 in measuring the effect of watershed development and the control of increased flow rates using a small detention structure. Table A-2 shows the peak discharges for the 10-year and 100-year design storms at the outlet to both subarea 1 and the total watershed; the peak rates are given for the before development, the after development, and the after development with detention conditions. It is evident that urbanization increases the 10-yr and 100-yr peak rates from subarea 1 from 75 to 139 cfs and from 204 to 308 cfs, respectively. The control structure that is located at the outlet of subarea 1 reduces the 10-yr and 100-yr discharges to 98 and 187 cfs, respectively. In this case, the control structure, which was defined in the STRUCT table, reduced the 100-yr event to a value below the before development peak discharge rate; the 10-yr rate increased even with the structure.

The TR-20 program is also useful for examining the effect of detention storage and development on downstream flow rates. Table A-2 also shows the peak rates at the outlet of the watershed for both the before development and after development with detention conditions. Both the 10-yr and 100-yr peak rates changed little with development because of the control structure located at the outlet of subarea 1. Thus, the control structure was able to mitigate the effects of development.

Summary

The purpose of this example was to show the card images that were necessary to solve a hypothetical, but realistic, example. It is evident that both the input requirements and the effort that would be required to formulate a solution would be minimal. This example does not include any of the control cards that would be required to access the TR-20 program; these were not discussed because they are very machine dependent, and it would be impractical to attempt to list all possibilities. I would expect that the user's in-house computer consultant can easily detail the control cards that are required to access the program.

TABLE A-2. Peak Discharges from Subarea 1 and at the Outlet

Site	Return Period	Before Development	After Development	After Development With Detention
Subarea 1	10-yrs	75 cfs	139 cfs	98 cfs
	100-yrs	204 cfs	308 cfs	187 cfs
Outlet	10-yrs	286 cfs	-	301 cfs
	100-yrs	722 cfs	-	698 cfs

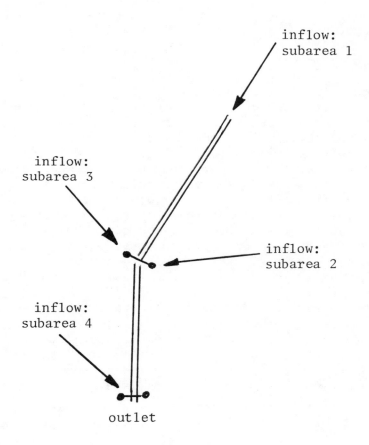

FIGURE A-1. Schematic Diagram of Watershed

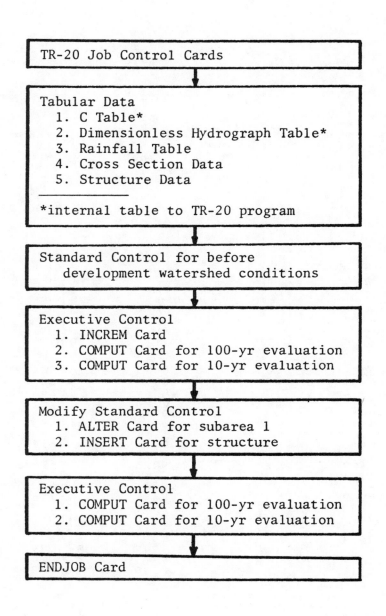

```
┌─────────────────────────────────────────────────┐
│ TR-20 Job Control Cards                         │
└─────────────────────────────────────────────────┘
                        │
                        ▼
┌─────────────────────────────────────────────────┐
│ Tabular Data                                    │
│     1.  C Table*                                │
│     2.  Dimensionless Hydrograph Table*         │
│     3.  Rainfall Table                          │
│     4.  Cross Section Data                      │
│     5.  Structure Data                          │
│     ─────────────                               │
│ *internal table to TR-20 program               │
└─────────────────────────────────────────────────┘
                        │
                        ▼
┌─────────────────────────────────────────────────┐
│ Standard Control for before                     │
│     development watershed conditions            │
└─────────────────────────────────────────────────┘
                        │
                        ▼
┌─────────────────────────────────────────────────┐
│ Executive Control                               │
│     1.  INCREM Card                             │
│     2.  COMPUT Card for 100-yr evaluation       │
│     3.  COMPUT Card for 10-yr evaluation        │
└─────────────────────────────────────────────────┘
                        │
                        ▼
┌─────────────────────────────────────────────────┐
│ Modify Standard Control                         │
│     1.  ALTER Card for subarea 1                │
│     2.  INSERT Card for structure               │
└─────────────────────────────────────────────────┘
                        │
                        ▼
┌─────────────────────────────────────────────────┐
│ Executive Control                               │
│     1.  COMPUT Card for 100-yr evaluation       │
│     2.  COMPUT Card for 10-yr evaluation        │
└─────────────────────────────────────────────────┘
                        │
                        ▼
┌─────────────────────────────────────────────────┐
│ ENDJOB Card                                     │
└─────────────────────────────────────────────────┘
```

FIGURE A-2. Flowchart of Input Data

```
JOB        TR-20                    FULLPRINT              MCCUEN
TITLE      12/1/80
  1 CTABLE                          0.20
  8                .0000     .0800        .1800      .2500      .3200
  8                .3700     .4100        .4500      .4900      .5100
  8                .5400     .5700        .5900      .6100      .6300
  8                .6500     .6600        .6700      .6900      .7000
  8                .7100     .7200        .7300      .7400      .7500
  8                .7600     .7700        .7700      .7800      .7900
  8                .7900     .8000        .8100      .8100      .8200
  8                .8200     .8300        .8300      .8400      .8400
  8                .8400     .8500        .8500      .8600      .8600
  8                .8600     .8600        .8700      .8700      .8700
  8                .8300     .3800        .8800      .8900      .8900
  8                .8900     .8900        .8900      .8900      .9000
  8                .9000     .9000        .9000      .9000      .9100
  8                .9100     .9100        .9100      .9100      .9100
  8                .9200     .9200        .9200      .9200      .9200
  9 ENDTBL        .9200     .9200        .9200      .9300      .9300
  4 DIMHYD                           0.02
  8      .0000     .0300        .1000      .1900      .3100
  8      .4700     .6600        .8200      .9300      .9900
  8    1.0000     .9900        .9300      .8600      .7800
  8      .6800     .5600        .4600      .3900      .3300
  8      .2800     .2410        .2070      .1740      .1470
  8      .1260     .1070        .0910      .0770      .0660
  8      .0550     .0470        .0400      .0340      .0290
  8      .0250     .0210        .0180      .0150      .0130
  8      .0110     .0090        .0080      .0070      .0060
  8      .0050     .0040        .0030      .0020      .0010
  8      .0000     .0000        .0000      .0000      .0000
  9 ENDTBL
  5 RAINFL 1                        .25
  8                .0000     .0027        .0053      .0081      .0108
  8                .0137     .0166        .0195      .0225      .0255
  8                .0286     .0318        .0350      .0383      .0417
  8                .0452     .0487        .0523      .0560      .0598
  8                .0637     .0677        .0718      .0761      .0804
  8                .0850     .0896        .0945      .0995      .1048
  8                .1101     .1159        .1217      .1280      .1345
  8                .1416     .1488        .1568      .1651      .1745
  8                .1842     .1956        .2076      .2223      .2384
  8                .2607     .2866        .4023      .6665      .7070
  8                .7384     .7580        .7757      .7894      .8022
  8                .8129     .8232        .8321      .8407      .8484
  8                .8559     .8628        .8694      .8756      .8816
  8                .8872     .8927        .8978      .9029      .9077
  8                .9124     .9169        .9213      .9255      .9297
  8                .9337     .9376        .9414      .9451      .9487
  8                .9523     .9557        .9591      .9624      .9656
  8                .9688     .9719        .9749      .9779      .9808
  8                .9887     .9865        .9893      .9920      .9948
  9 ENDTBL        .9974    1.0000       1.0000     1.0000     1.0000
  2 XSECTN    001                   1.
  8                          50.         0.         0.
  8                          53.        75.        30.
  8                          60.      2500.       600.
  9 ENDTBL
  2 XSECTN    002                   1.
  8                          50.         0.         0.
  8                          53.        75.        30.
  8                          60.      2500.       600.
  9 ENDTBL
  2 XSECTN    003                   1.
  8                          50.         0.         0.
```

FIGURE A-3.   Listing of Input for TR-20 Run

```
8                      53.         75.          30.
8                      60.       2500.         600.
9  ENDTBL
2  XSECTN   004         1.
8                      44.          0.           0.
8                      46.         90.          25.
8                      52.       2800.         450.
9  ENDTBL
2  XSECTN   005         1.
8                      44.          0.           0.
8                      46.         90.          25.
8                      52.       2800.         450.
9  ENDTBL
3  STRUCT       01
8                       0.          0.           0.
8                       1.         45.    0.573921
8                       2.         75.     1.37741
8                       2.        100.      2.2957
8                       4.        130.     3.67309
8                       5.        165.     5.28007
8                       6.        200.      7.3462
8                       7.        235.     9.64187
9  ENDTBL
6  RUNOFF 1 001      5  .1563     66.          .7     1 1 1
6  RUNOFF 1 002      6  .1250     71.          .6     1
6  REACH  3 003   5  7  1600.                         1
6  ADDHYL 4 003   6 7 5                               1
6  RUNOFF 1 003      6  .1719     63.          .9     1
6  ADDHYD 4 003   5 6 7                               1
6  RUNOFF 1 004      5  .1172     81.          .4     1
6  REACH  3 005   7  6  1150.                         1
6  ADDHYD 4 005   6 5 7                               1
   ENDATA
7  LIST
7  INCREM 6             .1
7  COMPUT 7 001  005    .0        7.          1.              1    2
   ENDCMP 1
7  COMPUT 7 001  005   0.         4.3         1.              1    2
   ENDCMP 1
7  ALTER  3
6  RUNOFF 1 001      5  .1563     75.          .55    1 1 1 1
7  INSERT 2 002
6  RESVOR 2      C1 5  7  0.0                          1 1 1 1
6  SAVMOV 5      01 7  5
7  LIST
7  COMPUT 7 001  005   0.         7.          1.              1    2
   ENDCMP 1
7  COMPUT 7 001  005   0.         4.3         1.              1    2
   ENDCMP 1
   ENDJOB 2
```

```
EXECUTIVE CONTROL CARD            OPERATION  LIST                                              PASS=  1

LISTING OF DATA IN CORE

    0

                      VELOCITY INCREMENT
1 CTABLE                       .2000
8                .0000         .0800         .1800         .2500         .3200
8                .3700         .4100         .4500         .4900         .5100
8                .5400         .5700         .5900         .6100         .6300
8                .6500         .6600         .6700         .6900         .7000
8                .7100         .7200         .7300         .7400         .7500
8                .7600         .7700         .7700         .7800         .7900
8                .7900         .8000         .8100         .8100         .8200
8                .8200         .8300         .8300         .8400         .8400
8                .8400         .8500         .8500         .8600         .8600
8                .8600         .8600         .8700         .8700         .8700
8                .8800         .8800         .8800         .8900         .8900
8                .8900         .8900         .8900         .8900         .9000
8                .9000         .9000         .9000         .9000         .9100
8                .9100         .9100         .9100         .9100         .9100
8                .9200         .9200         .9200         .9200         .9200
8                .9200         .9200         .9200         .9300         .9300
9 ENDTBL

          XSECTN NO.       DRAINAGE AREA
2 XSECTN      1                 1.0000
                            ELEVATION     DISCHARGE     END AREA
8                            50.0000         .0000         .0000
8                            53.0000       75.0000       30.0000
8                            60.0000     2500.0000      600.0000
9 ENDTBL

          XSECTN NO.       DRAINAGE AREA
2 XSECTN      2                 1.0000
                            ELEVATION     DISCHARGE     END AREA
8                            50.0000         .0000         .0000
8                            53.0000       75.0000       30.0000
8                            60.0000     2500.0000      600.0000
9 ENDTBL

          XSECTN NO.       DRAINAGE AREA
2 XSECTN      3                 1.0000
                            ELEVATION     DISCHARGE     END AREA
8                            50.0000         .0000         .0000
8                            53.0000       75.0000       30.0000
8                            60.0000     2500.0000      600.0000
9 ENDTBL

          XSECTN NO.       DRAINAGE AREA
2 XSECTN      4                 1.0000
                            ELEVATION     DISCHARGE     END AREA
8                            44.0000         .0000         .0000
8                            46.0000       90.0000       25.0000
8                            52.0000     2800.0000      450.0000
9 ENDTBL

          XSECTN NO.       DRAINAGE AREA
```

FIGURE A-4. Listing of Computer Output for TR-20 Run

```
2 XSECTN      5                1.0000
                        ELEVATION    DISCHARGE    END AREA
8                         44.0000       .0000        .0000
8                         46.0000     90.0000     25.0000
8                         52.0000   2800.0000    450.0000
9 ENDTBL

               STRUCT NO.
3 STRUCT          1
                        ELEVATION    DISCHARGE    STORAGE
8                          .0000        .0000        .0000
8                         1.0000     45.0000       .5739
8                         2.0000     75.0000      1.3774
8                         2.0000    100.0000      2.2957
8                         4.0000    130.0000      3.6731
8                         5.0000    165.0000      5.2801
8                         6.0000    200.0000      7.3462
8                         7.0000    235.0000      9.6419
9 ENDTBL

                  TIME INCREMENT
4 DIMHYD               .0000
8            .0000        .0300        .1000        .1900        .3100
8            .4700        .6600        .8200        .9300        .9900
8           1.0000        .9900        .9300        .8600        .7800
8            .6800        .5600        .4600        .3900        .3300
8            .2800        .2410        .2070        .1740        .1470
8            .1260        .1070        .0910        .0770        .0660
8            .0550        .0470        .0400        .0340        .0290
8            .0250        .0210        .0180        .0150        .0130
8            .0110        .0090        .0080        .0070        .0060
8            .0050        .0040        .0030        .0020        .0010
8            .0000        .0000        .0000        .0000        .0000
9 ENDTBL
  COMPUTED PEAK K FACTOR =    484.00


                  TIME INCREMENT
5 RAINFL 1             .2500
8            .0000        .0027        .0053        .0081        .0108
8            .0137        .0166        .0195        .0225        .0255
8            .0286        .0318        .0350        .0383        .0417
8            .0452        .0487        .0523        .0560        .0598
8            .0637        .0677        .0718        .0761        .0804
8            .0850        .0896        .0945        .0995        .1048
8            .1101        .1159        .1217        .1280        .1345
8            .1416        .1488        .1568        .1651        .1745
8            .1842        .1956        .2076        .2223        .2384
8            .2607        .2866        .4023        .6665        .7070
8            .7384        .7580        .7757        .7894        .8022
8            .8129        .8232        .8321        .8407        .8484
8            .8559        .8628        .8694        .8756        .8816
8            .8872        .8927        .8978        .9029        .9077
8            .9124        .9169        .9213        .9255        .9297
8            .9337        .9376        .9414        .9451        .9487
8            .9523        .9557        .9591        .9624        .9656
8            .9688        .9719        .9749        .9779        .9808
8            .9887        .9865        .9893        .9920        .9948
8            .9974       1.0000       1.0000       1.0000       1.0000
9 ENDTBL
```

133

STANDARD CONTROL INSTRUCTIONS

```
6 RUNOFF 1   1       5          .1563      66.0000      .70001 1 1 1 0  0
6 RUNOFF 1   2       6          .1250      71.0000      .60001 0 0 0 0  0
6 REACH  3   3   5   7     1600.0000        .0000       .00001 0 0 0 0  0
6 ADDHYD 4   3   6 7 5                                       1 0 0 0 0  0
6 RUNOFF 1   3       6          .1719      63.0000      .90001 0 0 0 0  0
6 ADDHYD 4   3   5 6 7                                       1 0 0 0 0  0
6 RUNOFF 1   4       5          .1172      81.0000      .40001 0 0 0 0  0
6 REACH  3   5   7   6     1150.0000        .0000       .00001 0 0 0 0  0
6 ADDHYD 4   5   6 5 7                                       1 0 0 0 0  0
  ENDATA

END OF LISTING
```

134

EXECUTIVE CONTROL CARD
EXECUTIVE CONTROL CARD

OPERATION INCREM.
OPERATION COMPUT.
RAIN DEPTH= 7.00
STARTING TIME= .00
ALTERNATE NO.= 0
STORM NO.= 0

MAIN TIME INCREMENT= .10
FROM XSECTN/STRUCT. 1/ 0   TO XSECTN/STRUCT 5/ 0
RAIN DURATION= 1.00   RAIN TABLE NO.= 1   SOIL CONDITION= 2

PASS= 1

SUBROUTINE RUNOFF   CROSS SECTION 1
AREA= .16   INPUT RUNOFF CURVE= 66.0   TIME OF CONCENTRATION= .7C

PEAK TIMES	PEAK DISCHARGES	PEAK ELEVATIONS	DRAINAGE AREA= .16
12.32	203.638	(RUNOFF)	
22.70	9.219	(RUNOFF)	
23.95	5.848	(RUNOFF)	

HYDROGRAPH, TZERO= .00   DELTA T= .10

TIME						
.00	DISCHG	50.00	50.00	50.00	50.00	50.00
	ELEV	50.00	50.00	50.00	50.00	50.00
1.00	DISCHG	50.00	50.00	50.00	50.00	50.00
	ELEV	50.00	50.00	50.00	50.00	50.00
1.00	DISCHG	50.00	50.00	50.00	50.00	50.00
	ELEV	50.00	50.00	50.00	50.00	50.00
2.00	DISCHG	50.00	50.00	50.00	50.00	50.00
	ELEV	50.00	50.00	50.00	50.00	50.00
3.00	DISCHG	50.00	50.00	50.00	50.00	50.00
	ELEV	50.00	50.00	50.00	50.00	50.00
3.00	DISCHG	50.00	50.00	50.00	50.00	50.00
	ELEV	50.00	50.00	50.00	50.00	50.00
4.00	DISCHG	50.00	50.00	50.00	50.00	50.00
	ELEV	50.00	50.00	50.00	50.00	50.00
4.00	DISCHG	50.00	50.00	50.00	50.00	50.00
	ELEV	50.00	50.00	50.00	50.00	50.00
5.00	DISCHG	50.00	50.00	50.00	50.00	50.00
	ELEV	50.00	50.00	50.00	50.00	50.00
5.00	DISCHG	50.00	50.00	50.00	50.00	50.00
	ELEV	50.00	50.00	50.00	50.00	50.00
6.00	DISCHG	50.00	50.00	50.00	50.00	50.00
	ELEV	50.00	50.00	50.00	50.00	50.00
7.00	DISCHG	50.00	50.00	50.00	50.00	50.00
	ELEV	50.00	50.00	50.00	50.00	50.00
8.00	DISCHG	50.00	50.00	50.00	50.00	50.00
	ELEV	50.00	50.00	50.00	50.00	50.92
9.00	DISCHG	50.01	50.07	50.13	50.23	50.36
	ELEV	50.00	50.03	50.71	50.02	50.04
10.00	DISCHG	1.40	50.08	52.35	50.11	53.03
	ELEV	1.15	50.08	52.09	52.11	50.15
11.00	DISCHG	6.31	8.45	9.86	11.80	38.79
	ELEV	5.49	7.28	50.32	15.80	23.69
12.00	DISCHG	104.26	203.29	196.52	146.94	100.35
	ELEV	152.30	188.86	175.36	120.37	84.59
13.00	DISCHG	71.86	53.37	38.19	27.03	21.11
	ELEV	53.08	47.27	42.27	34.82	29.77
14.00	DISCHG	52.87	51.89	20.46	51.39	17.59
	ELEV	41.58	51.35	51.53	51.28	21.25
15.00	DISCHG	26.07	23.26	50.85	50.82	13.95
	ELEV	24.57	22.16	50.93	50.76	14.25
16.00	DISCHG	17.37	16.03	14.89	50.79	11.63
	ELEV	16.90	50.64	15.25	50.57	11.47
17.00	DISCHG	13.66	12.90	12.56	50.63	10.47
	ELEV	13.40	12.56	12.44	50.58	10.21
18.00	DISCHG	50.54	50.52	50.50	50.49	10.07
	ELEV	11.28	50.63	12.00	50.48	10.41
19.00	DISCHG	11.11	50.44	50.43	50.42	50.40
	ELEV	50.46	10.52	10.51	50.43	50.36
20.00	DISCHG	9.82	9.70	9.58	9.47	9.35
	ELEV	9.94	9.58	50.37	50.38	8.07
21.00	DISCHG	8.75	8.66	8.57	8.48	8.31
	ELEV	8.84	50.35	50.37	50.34	8.15
22.00	DISCHG	7.92	7.85	7.78	57.64	50.32
	ELEV	57.99	50.33	50.31	57.71	7.39
23.00	DISCHG	57.32	57.29	57.31	57.03	50.30
	ELEV	57.26	50.31	57.14	57.03	50.27
24.00	DISCHG	56.77	50.29	50.28	56.98	6.82
	ELEV	56.72	50.27	57.28	56.79	6.68
25.00	DISCHG	56.27	56.27	56.21	57.24	50.31
	ELEV	56.77	50.27	57.98	58.79	50.31
	DISCHG	56.51	55.65	55.41	55.60	50.83
	ELEV	56.51	50.27	50.22	55.77	50.81
	DISCHG	55.83	55.32	55.22	50.23	50.76
	ELEV	55.26	50.23	52.93	50.23	50.76
	DISCHG	50.23	50.21	54.66	52.08	51.06
	ELEV	50.54	50.21	50.39	50.12	51.04
	DISCHG	50.02	50.02	50.19	50.15	50.03
	ELEV	50.02	50.02	50.39	50.09	50.03
26.00	DISCHG	50.01	50.00	50.01	50.00	50.00
	ELEV	50.00	50.00	50.01	50.00	50.00

TOTAL WATER, IN INCHES ON DRAINAGE AREA= 3.2027   CFS-HRS= 323.06   ACRE-FT= 26.70

SUBROUTINE RUNOFF   CROSS SECTION 2
AREA= .13   INPUT RUNOFF CURVE= 71.0   TIME OF CONCENTRATION= .60

135

PEAK TIMES | PEAK DISCHARGES | PEAK ELEVATIONS
12.25 | 208.516 | (RUNOFF)
22.67 | 8.559 | (RUNOFF)
23.79 | 5.045 | (RUNOFF)

SUBROUTINE REACH   CROSS SECTION 3   INPUT COEFFICIENT= .0000   INPUT ROUTINGS= .00
LENGTH= 1600.00
AVERAGE WATER VELOCITY= 3.177   AVERAGE ROUTING COEFF= .6514   NUMBER OF ROUTINGS= .91

PEAK TIMES | PEAK DISCHARGES | PEAK ELEVATIONS
12.46 | 195.386 | 53.35
22.83 | 8.974 | 50.36
24.05 | 5.830 | 50.23

SUBROUTINE ADDHYD   CROSS SECTION 3   OUTPUT HYDROGRAPH= 5
INPUT HYDROGRAPHS= 6,7

PEAK TIMES | PEAK DISCHARGES | PEAK ELEVATIONS
12.34 | 381.850 | 53.89
22.73 | 17.038 | 50.68
23.92 | 10.824 | 50.43

SUBROUTINE RUNOFF   CROSS SECTION 3   TIME OF CONCENTRATION= .90
AREA= .17   INPUT RUNOFF CURVE= 63.0

PEAK TIMES | PEAK DISCHARGES | PEAK ELEVATIONS
12.47 | 169.322 | (RUNOFF)
22.78 | 8.974 | (RUNOFF)
23.88 | 6.110 | (RUNOFF)

SUBROUTINE ADDHYD   CROSS SECTION 3   OUTPUT HYDROGRAPH= 7
INPUT HYDROGRAPHS= 5,6

PEAK TIMES | PEAK DISCHARGES | PEAK ELEVATIONS
12.38 | 542.289 | 54.35
22.74 | 25.960 | 51.04
23.90 | 16.932 | 50.68

SUBROUTINE RUNOFF   CROSS SECTION 4   TIME OF CONCENTRATION= .40
AREA= .12   INPUT RUNOFF CURVE= 81.0

PEAK TIMES | PEAK DISCHARGES | PEAK ELEVATIONS
12.12 | 360.247 |
22.59 | 10.780 |
23.61 | 5.260 |

SUBROUTINE REACH   CROSS SECTION 5   INPUT COEFFICIENT= .0000   INPUT ROUTINGS= .00
LENGTH= 1150.00
AVERAGE WATER VELOCITY= 5.465   AVERAGE ROUTING COEFF= .7627   NUMBER OF ROUTINGS= .45

PEAK TIMES | PEAK DISCHARGES | PEAK ELEVATIONS
12.43 | 536.289 | 46.99
22.79 | 25.693 | 44.57
23.95 | 16.921 | 44.38

SUBROUTINE ADDHYD   CROSS SECTION 5   OUTPUT HYDROGRAPH= 7
INPUT HYDROGRAPHS= 6,5

PEAK TIMES | PEAK DISCHARGES | PEAK ELEVATIONS
12.27 | 722.138 | 47.40
22.66 | 34.465 | 44.77
23.93 | 22.106 | 44.49

ENDCMP

EXECUTIVE CONTROL CARD
  STARTING TIME=   .00   RAIN DEPTH= 4.30   RAIN DURATION= 1.00   RAIN TABLE NO.= 1   SOIL CONDITION= 2
  ALTERNATE NO.= 0   STORM NO.= 0

SUBROUTINE RUNOFF   CROSS SECTION 1
  AREA=   .16   INPUT RUNOFF CURVE= 66.0   TIME OF CONCENTRATION=   .70

             PEAK DISCHARGES   PEAK ELEVATIONS
  PEAK TIMES   74.503   (RUNOFF)
   12.35   4.489   (RUNOFF)
   22.70   2.857   (RUNOFF)
   23.95

                          HYDROGRAPH, TZERO=   .00   DELTA T= .10   DRAINAGE AREA=   .16

TIME
  .00   DISCHG   50.00   50.00   50.00   50.00   50.00   50.00   50.00
        ELEV   50.00   50.00   50.00   50.00   50.00   50.00   50.00
 1.00   DISCHG   50.00   50.00   50.00   50.00   50.00   50.00   50.00
        ELEV   50.00   50.00   50.00   50.00   50.00   50.00   50.00
 2.00   DISCHG   50.00   50.00   50.00   50.00   50.00   50.00   50.00
        ELEV   50.00   50.00   50.00   50.00   50.00   50.00   50.00
 3.00   DISCHG   50.00   50.00   50.00   50.00   50.00   50.00   50.00
        ELEV   50.00   50.00   50.00   50.00   50.00   50.00   50.00
 4.00   DISCHG   50.00   50.00   50.00   50.00   50.00   50.00   50.00
        ELEV   50.00   50.00   50.00   50.00   50.00   50.00   50.00
 5.00   DISCHG   50.00   50.00   50.00   50.00   50.00   50.00   50.00
        ELEV   50.00   50.00   50.00   50.00   50.00   50.00   50.00
 6.00   DISCHG   50.00   50.00   50.00   50.00   50.00   50.00   50.00
        ELEV   50.00   50.00   50.00   50.00   50.00   50.00   50.00
 7.00   DISCHG   50.00   50.00   50.00   50.00   50.00   50.00   50.00
        ELEV   50.00   50.00   50.00   50.00   50.00   50.00   50.00
 8.00   DISCHG   50.00   50.00   50.00   50.00   50.00   50.00   50.00
        ELEV   50.00   50.00   50.00   50.00   50.00   50.00   50.00
 9.00   DISCHG   50.00   50.00   50.00   50.00   50.00   50.00   50.00
        ELEV   50.00   50.00   50.00   50.00   50.00   50.00   50.00
10.00   DISCHG   50.04   50.01   50.11   50.25   50.57   50.06   16.04
        ELEV   50.61   49.77   52.95   73.31   67.03   51.42   57.77
11.00   DISCHG   30.61   51.98   52.93   73.63   52.44   52.41   40.31
        ELEV   29.90   25.98   20.42   18.44   52.81   10.45   47.81
12.00   DISCHG   51.20   50.71   50.82   50.74   59.43   50.62   13.35
        ELEV   51.84   11.21   10.17   59.78   50.57   14.31
13.00   DISCHG   50.47   50.43   50.17   50.29   50.67   50.62   58.30
        ELEV   8.07   57.66   57.47   57.29   50.38   50.11   58.50
14.00   DISCHG   50.31   50.30   50.47   50.12   50.36   50.37   56.55
        ELEV   6.42   50.25   57.28   50.82   50.33
15.00   DISCHG   50.26   50.24   50.08   50.29   50.27   50.28   50.26
        ELEV   50.35   50.25   50.24   50.87   50.27   55.22
16.00   DISCHG   50.22   50.21   50.24   50.24   50.22   50.23   54.80
        ELEV   54.42   54.92   55.23   54.86
17.00   DISCHG   50.21   50.21   50.20   50.20   50.20   50.20   54.19
        ELEV   54.74   54.63   54.58   54.52   54.20   54.17
18.00   DISCHG   50.19   50.19   50.18   50.18   50.18   50.18   53.88
        ELEV   54.24   54.31   54.18   54.07   54.03   53.96
19.00   DISCHG   50.17   50.17   50.16   50.16   50.16   50.16   53.57
        ELEV   53.85   53.78   53.75   53.72   53.99   53.92
20.00   DISCHG   50.15   50.14   50.15   50.15   50.15   50.15   53.11
        ELEV   53.52   53.49   53.65   53.36   53.65   53.34
21.00   DISCHG   50.14   50.13   50.13   50.14   50.13   50.13   53.74
        ELEV   53.29   53.24   53.46   53.43   53.38   53.13
22.00   DISCHG   50.13   50.13   50.13   50.14   50.51   50.17   54.18
        ELEV   3.17   3.27   3.29   3.88   54.28   54.30
23.00   DISCHG   52.75   52.55   52.54   52.04   52.79   52.82   52.85
        ELEV   50.17   50.15   50.16   50.73   52.82
24.00   DISCHG   52.78   50.60   50.23   50.11   50.11   50.73   50.37
        ELEV   50.11   50.11   1.87   1.43   1.03   50.52
25.00   DISCHG   50.19   50.13   50.09   50.09   50.06   50.04   50.01
        ELEV   50.26   50.11   50.07   50.05   50.03
26.00   DISCHG   50.01   50.01   50.00   50.00   50.00   50.03   50.00
        ELEV   50.00   50.00   50.00   50.00   50.00

  TOTAL WATER, IN INCHES ON DRAINAGE AREA=   1.2693   CFS-HRS= 128.04   ACRE-FT=   10.58

SUBROUTINE RUNOFF   CROSS SECTION 2
  AREA=   .13   INPUT RUNOFF CURVE= 71.0   TIME OF CONCENTRATION=   .60

```
                        PEAK TIMES           PEAK DISCHARGES        PEAK ELEVATIONS
                         12.27                   86.648               (RUNOFF)
                         22.67                    4.410               (RUNOFF)
                         23.83                    2.606               (RUNOFF)

SUBROUTINE REACH      CROSS SECTION 3
           LENGTH=  1600.00    INPUT COEFFICIENT=  .0000    INPUT ROUTINGS=   .00    NUMBER OF ROUTINGS=   1.06

           AVERAGE WATER VELOCITY=  2.500    AVERAGE ROUTING COEFF=  .5952

                        PEAK TIMES           PEAK DISCHARGES        PEAK ELEVATIONS
                         12.52                   70.306                52.81
                         22.85                    4.339                50.17
                         24.07                    2.845                50.11

SUBROUTINE ADDHYD    CROSS SECTION 3
           INPUT HYDROGRAPHS= 6,7       OUTPUT HYDROGRAPH= 5

                        PEAK TIMES           PEAK DISCHARGES        PEAK ELEVATIONS
                         12.38                  143.950                53.20
                         22.73                    8.434                50.34
                         23.93                    5.419                50.22

SUBROUTINE RUNOFF     CROSS SECTION 3       INPUT RUNOFF CURVE=  63.0     TIME OF CONCENTRATION=   .90
           AREA=      .17

                        PEAK TIMES           PEAK DISCHARGES        PEAK ELEVATIONS
                         12.51                   56.518               (RUNOFF)
                         22.78                    4.205               (RUNOFF)
                         23.88                    2.873               (RUNOFF)

SUBROUTINE ADDHYD    CROSS SECTION 3
           INPUT HYDROGRAPHS= 5,6       OUTPUT HYDROGRAPH= 7

                        PEAK TIMES           PEAK DISCHARGES        PEAK ELEVATIONS
                         12.42                  197.374                53.35
                         22.74                   12.615                50.50
                         23.91                    8.290                50.33

SUBROUTINE RUNOFF     CROSS SECTION 4       INPUT RUNOFF CURVE=  81.0     TIME OF CONCENTRATION=   .40
           AREA=      .12

                        PEAK TIMES           PEAK DISCHARGES        PEAK ELEVATIONS
                         12.13                  160.405                46.23
                         22.59                    6.080                44.28
                         23.61                    2.970                44.18

SUBROUTINE REACH      CROSS SECTION 5
           LENGTH=  1150.00    INPUT COEFFICIENT=  .0000    INPUT ROUTINGS=   .00    NUMBER OF ROUTINGS=    .53

           AVERAGE WATER VELOCITY=  4.345    AVERAGE ROUTING COEFF=  .7188

                        PEAK TIMES           PEAK DISCHARGES        PEAK ELEVATIONS
                         12.49                  194.138                46.43
                         22.81                   12.457                44.39
                         23.97                    8.281                44.25

SUBROUTINE ADDHYD    CROSS SECTION 5
           INPUT HYDROGRAPHS= 6,5       OUTPUT HYDROGRAPH= 7

                        PEAK TIMES           PEAK DISCHARGES
                         12.25                  286.082
                         22.65                   17.426
                         23.95                   11.237

ENDCMP
```

CHANGES TO STANDARD CONTROL LIST FOLLOW
EXECUTIVE CONTROL CARD   OPERATION ALTER

STANDARD CONTROL CARD, SUBROUTINE RUNOFF, CROSS-SECTION=   1 STRUCTURE=   0
IN1 HYD=0   IN2 HYD=5   OUT HYD=5   DATA FIELDS=   .1563   66.0000
OUTPUT OPTION= 1 1 1 0 OID=   OPERATION   INSERT.   XSECTN/STRUCT=   2/   0   .7000
EXECUTIVE CONTROL CARD

STANDARD CONTROL CARD, SUBROUTINE RESVOR, CROSS-SECTION=   0 STRUCTURE=   1
IN1 HYD=5   IN2 HYD=0   OUT HYD=7   DATA FIELDS=   .0000   .0000   .0000
OUTPUT OPTION= 1 1 1 1 0 OID=

STANDARD CONTROL CARD, SUBROUTINE SAVMOV, CROSS-SECTION=   0 STRUCTURE=   1
IN1 HYD=7   IN2 HYD=0   OUT HYD=0   DATA FIELDS=   .0000   .0000   .0000
OUTPUT OPTION= 0 0 0 0 0 OID=

EXECUTIVE CONTROL CARD   OPERATION INCREM,   MAIN TIME INCREMENT= 1/ .10
EXECUTIVE CONTROL CARD   OPERATION COMPUT,   FROM XSECTN/STRUCT 1/ 0   TO XSECTN/STRUCT 5/ 0
          STARTING TIME = .00   RAIN DEPTH= 7.00   RAIN DURATION= 1.00   RAIN TABLE NO.= 1   SOIL CONDITION= 2
          ALTERNATE NO.= 0   STORM NO.= 0

SUBROUTINE RUNOFF   CROSS SECTION 1
AREA= .16   INPUT RUNOFF CURVE= 75.0   TIME OF CONCENTRATION= .55

PEAK TIMES   PEAK DISCHARGES   PEAK ELEVATIONS
  12.22   308.275   (RUNOFF)
  22.66   11.607   (RUNOFF)
  23.76   6.600   (RUNOFF)

                                                        DRAINAGE AREA= .16

HYDROGRAPH, TZERO= .00   DELTA T= .10

TIME	DISCHG	ELEV				
.000	50.00	50.00	50.00	50.00	50.00	50.00
.000						
1.000						
...						
25.00						

TOTAL WATER, IN INCHES ON DRAINAGE AREA= 4.1457   CFS-HRS= 418.18   ACRE-FT= 34.56

SUBROUTINE RUNOFF   CROSS SECTION 2
AREA= .13   INPUT RUNOFF CURVE= 71.0   TIME OF CONCENTRATION= .60

PEAK TIMES   PEAK ELEVATIONS
PEAK DISCHARGES

12.25
22.67
23.79

208.516
8.559
5.045

(RUNOFF)
(RUNOFF)
(RUNOFF)

SUBROUTINE RESVOR STRUCTURE 1

SURFACE ELEVATION= .00

PEAK TIMES
12.52
22.76
23.92

PEAK DISCHARGES
167.208
10.616
6.571

PEAK ELEVATIONS
5.63
.24
.15

DELTA T= .10

HYDROGRAPH, TZERO= .00

DRAINAGE AREA= .16

TIME	DISCHG	ELEV				
.00	.00	.00	.00	.00	.00	.00
1.00	.00	.00	.00	.00	.00	.00
2.00	.00	.00	.00	.00	.00	.00
3.00	.00	.00	.00	.00	.00	.00
4.00	.00	.00	.00	.00	.00	.00
5.00	.00	.00	.00	.00	.00	.00
6.00	.00	.00	.00	.00	.00	.00

TOTAL WATER, IN INCHES ON DRAINAGE AREA= 4.1452   CFS-HRS=   418.13   ACRE-FT=   34.55

SUBROUTINE SAVMOV STRUCTURE 1

INPUT HYDROGRAPH= 7   OUTPUT HYDROGRAPH= 5

SUBROUTINE REACH   CROSS SECTION 3
LENGTH= 1600.00   INPUT COEFFICIENT=   .0000   INPUT ROUTINGS=   .00

AVERAGE WATER VELOCITY= 3.115   AVERAGE ROUTING COEFF= .6469   NUMBER OF ROUTINGS=   .92

PEAK TIMES   PEAK DISCHARGES   PEAK ELEVATIONS
  12.67   184.134   53.32
  22.89   10.324   50.41
  24.05   6.564   50.26

SUBROUTINE ADDHYD   CROSS SECTION 3
INPUT HYDROGRAPHS= 6,7   OUTPUT HYDROGRAPH= 5

PEAK TIMES   PEAK DISCHARGES   PEAK ELEVATIONS
  12.36   342.165   53.77
  22.76   17.628   50.71
  23.92   11.563   50.46

SUBROUTINE RUNOFF   CROSS SECTION 3
AREA=   .17   INPUT RUNOFF CURVE= 63.0   TIME OF CONCENTRATION= .90

PEAK TIMES   PEAK DISCHARGES   PEAK ELEVATIONS
  12.47   169.322   (RUNOFF)
  22.78   8.974   (RUNOFF)
  23.88   6.110   (RUNOFF)

SUBROUTINE ADDHYD   CROSS SECTION 5,6
INPUT HYDROGRAPHS= 5,6   OUTPUT HYDROGRAPH= 7

PEAK TIMES   PEAK DISCHARGES   PEAK ELEVATIONS
  12.40   506.423   54.25
  22.76   26.797   51.07
  23.90   17.671   50.71

SUBROUTINE RUNOFF   CROSS SECTION 4
AREA=   .12   INPUT RUNOFF CURVE= 81.0   TIME OF CONCENTRATION= .40

PEAK TIMES   PEAK DISCHARGES   PEAK ELEVATIONS
  12.12   320.247   (RUNOFF)
  22.59   10.780   (RUNOFF)
  23.61   5.260   (RUNOFF)

SUBROUTINE REACH   CROSS SECTION 5
LENGTH= 1150.00   INPUT COEFFICIENT=   .0000   INPUT ROUTINGS=   .00

AVERAGE WATER VELOCITY= 5.389   AVERAGE ROUTING COEFF= .7602   NUMBER OF ROUTINGS=   .45

PEAK TIMES   PEAK DISCHARGES   PEAK ELEVATIONS
  12.46   501.332   46.91
  22.81   26.540   44.59
  23.95   17.660   44.39

SUBROUTINE ADDHYD   CROSS SECTION 6,5
INPUT HYDROGRAPHS= 6,5   OUTPUT HYDROGRAPH= 7

PEAK TIMES   PEAK DISCHARGES   PEAK ELEVATIONS
  12.24   698.150   47.35
  22.66   34.979   44.78
  23.93   22.644   44.51

ENDCMP

142

EXECUTIVE CONTROL CARD   OPERATION COMPUT.   FROM XSECTN/STRUCT 1/ 0   TO XSECTN/STRUCT 1/ 0
STARTING TIME= .00   RAIN DEPTH= 4.30   RAIN DURATION= 1.00   RAIN TABLE NO.= 1   SOIL CONDITION= 2
ALTERNATE NO.= 0   STORM NO.= 0

SUBROUTINE RUNOFF   CROSS SECTION 1
AREA= .16   INPUT RUNOFF CURVE= 75.0   TIME OF CONCENTRATION= .55

PEAK DISCHARGES   PEAK ELEVATIONS
138.774   (RUNOFF)
6.220   (RUNOFF)
3.544   (RUNOFF)

PEAK TIMES
12.23
22.66
23.77

HYDROGRAPH, TZERO= .00   DELTA T= .10   DRAINAGE AREA= .16

TIME						
.00	DISCHG	50.00	50.00	50.00	50.00	50.00
.00	ELEV	50.00	50.00	50.00	50.00	50.00
1.00	DISCHG	50.00	50.00	50.00	50.00	50.00
1.00	ELEV	50.00	50.00	50.00	50.00	50.00
2.00	DISCHG	50.00	50.00	50.00	50.00	50.00
2.00	ELEV	50.00	50.00	50.00	50.00	50.00
3.00	DISCHG	50.00	50.00	50.00	50.00	50.00
3.00	ELEV	50.00	50.00	50.00	50.00	50.00
4.00	DISCHG	50.00	50.00	50.00	50.00	50.00
4.00	ELEV	50.00	50.00	50.00	50.00	50.00
5.00	DISCHG	50.00	50.00	50.00	50.00	50.00
5.00	ELEV	50.00	50.00	50.00	50.00	50.00
6.00	DISCHG	50.00	50.00	50.00	50.00	50.00
6.00	ELEV	50.00	50.00	50.00	50.00	50.00
7.00	DISCHG	50.00	50.00	50.00	50.00	50.00
7.00	ELEV	50.00	50.00	50.00	50.00	50.00
8.00	DISCHG	50.00	50.00	50.00	50.00	50.00
8.00	ELEV	50.00	50.00	50.00	50.00	50.00
9.00	DISCHG	50.00	50.00	50.00	50.01	50.09
9.00	ELEV	50.04	50.01	50.17	50.28	50.40
10.00	DISCHG	50.54	50.69	51.29	51.52	52.45
10.00	ELEV	51.79	51.06	52.08	52.95	52.31
11.00	DISCHG	56.29	54.46	55.25	56.25	57.40
11.00	ELEV	57.67	51.57	51.89	52.00	50.30
12.00	DISCHG	86.03	120.15	137.80	113.86	71.56
12.00	ELEV	53.11	52.86	52.79	52.12	51.63
13.00	DISCHG	53.33	32.92	29.04	20.18	14.81
13.00	ELEV	19.17	12.92	10.02	50.59	50.31
14.00	DISCHG	11.12	13.03	12.48	11.07	50.62
14.00	ELEV	10.96	50.45	50.47	50.42	10.31
15.00	DISCHG	50.52	50.54	50.51	50.48	50.44
15.00	ELEV	50.35	50.87	50.41	50.35	50.16
16.00	DISCHG	10.62	50.31	50.36	50.08	50.33
16.00	ELEV	57.39	57.43	57.31	57.09	56.85
17.00	DISCHG	50.32	50.72	57.57	57.20	56.97
17.00	ELEV	50.28	50.46	50.37	50.28	56.27
18.00	DISCHG	50.75	50.26	50.37	50.25	50.24
18.00	ELEV	50.67	50.68	50.64	50.12	50.04
19.00	DISCHG	50.23	50.09	50.09	50.14	50.37
19.00	ELEV	55.89	55.11	55.61	55.22	50.21
20.00	DISCHG	50.21	50.15	50.04	50.22	50.22
20.00	ELEV	50.76	54.63	54.99	54.90	54.80
21.00	DISCHG	54.19	54.61	54.29	54.25	54.19
21.00	ELEV	54.37	54.30	54.26	54.18	54.07
22.00	DISCHG	54.17	54.17	54.18	54.18	54.16
22.00	ELEV	54.02	53.99	54.07	54.16	54.16
23.00	DISCHG	53.16	53.13	50.18	50.24	50.17
23.00	ELEV	53.13	53.35	53.49	53.54	53.54
24.00	DISCHG	53.04	52.98	50.13	50.14	50.14
24.00	ELEV	52.12	50.14	50.15	50.46	50.20
25.00	DISCHG	50.01	50.05	50.03	50.01	50.01
25.00	ELEV	50.00	50.00	50.00	50.00	50.01

TOTAL WATER, IN INCHES ON DRAINAGE AREA= 1.8940   CFS-HRS= 191.05   ACRE-FT= 15.79

SUBROUTINE RUNOFF   CROSS SECTION 2
AREA= .13   INPUT RUNOFF CURVE= 71.0   TIME OF CONCENTRATION= .60

PEAK DISCHARGES   PEAK ELEVATIONS
86.648   (RUNOFF)

PEAK TIMES
12.27

SUBROUTINE RESVOR STRUCTURE   1   .00

			(RUNOFF)
22.67	4.410		(RUNOFF)
23.83	2.606		

SURFACE ELEVATION=

PEAK DISCHARGES   HYDROGRAPH TZERO= .00   PEAK ELEVATIONS   DELTA T= .10   DRAINAGE AREA= .16

PEAK TIMES   98.180   2.00
12.47   5.669   2.13
22.76   3.531   .08
23.94

TIME													
.00	DISCHG	.00	.00	.00	.00	.00	.00	.00	.00	.00	.00	.00	.00
	ELEV	.00	.00	.00	.00	.00	.00	.00	.00	.00			
1.00	DISCHG	.00	.00	.00	.00	.00	.00	.00	.00	.00			
	ELEV												

(Numerical hydrograph output table — DISCHG / ELEV values for TIME = .00 through 25.00, continued across multiple discharge and elevation columns.)

Peak discharge values in columns include: 3.07, 47.07, 63.30, 15.61, 10.50, 8.26, 6.92, 6.01, 5.34, 5.12, 4.83, 4.42, 4.39, 4.06, 4.09, 3.83, 3.52, 3.48, 3.08, .33, .01

...

47.40, 14.01, 90.00, 62.06, 54.71, 14.71, 10.33, 8.23, 6.18, 6.15, 5.93, 5.28, 4.78, 4.39, 4.06, 3.85, 3.48, 3.32, 3.07, .22, .10

Column group values: 96.95, 30.23, 12.67, 9.29, 7.66, 6.22, 5.72, 5.13, 5.07, 4.61, 4.28, 4.14, 4.09, 3.35, 2.54, .06, .00

98.03, 25.24, 12.56, 9.27, 7.52, 6.42, 5.65, 5.13, 4.11, 4.06, 4.10, 4.15, 5.65, 3.52, .73, .01, .00

94.61, 21.93, 11.93, 9.01, 7.39, 6.31, 5.59, 5.12, 5.02, 4.57, 4.21, 4.09, 5.12, 3.11, 1.49, .02, .00

88.56, 19.02, 11.54, 8.61, 7.27, 6.16, 5.53, 5.14, 4.97, 4.54, 4.18, 4.09, 5.12, 3.46, 1.06, .02, .01, .00

73.15, 16.59, 10.83, 8.44, 7.03, 6.08, 5.40, 4.87, 4.46, 4.12, 4.10, 5.20, 3.53, .50, .01, .00

TOTAL WATER, IN INCHES ON DRAINAGE AREA=   1.8931   190.96   CFS-HRS=   ACRE-FT=   15.78

SUBROUTINE SAVMOV STRUCTURE   1   OUTPUT HYDROGRAPH= 5
INPUT HYDROGRAPH= 7

SUBROUTINE REACH   CROSS SECTION 3
LENGTH= 1800.00   INPUT COEFFICIENT=   .0000   INPUT ROUTINGS=   .00

144

```
                    AVERAGE WATER VELOCITY=    2.603    AVERAGE ROUTING COEFF=   .6050    NUMBER OF ROUTINGS=   1.03

                 PEAK TIMES        PEAK DISCHARGES        PEAK ELEVATIONS
                  12.65              95.566                 53.06
                  22.91               5.497                 50.22
                  24.07               3.522                 50.14

SUBROUTINE ADDHYD    CROSS SECTION                  3   OUTPUT HYDROGRAPH=  5
           INPUT HYDROGRAPHS=  6,7

                 PEAK TIMES        PEAK DISCHARGES        PEAK ELEVATIONS
                  12.41             159.590                 53.24
                  22.76               9.278                 50.37
                  23.94               8.106                 50.24

SUBROUTINE RUNOFF    CROSS SECTION  3  INPUT RUNOFF CURVE=  63.0    TIME OF CONCENTRATION=   .90
           AREA=    .17

                 PEAK TIMES        PEAK DISCHARGES        PEAK ELEVATIONS
                  12.51              56.518                (RUNOFF)
                  22.78               4.205                (RUNOFF)
                  23.88               2.673                (RUNOFF)

SUBROUTINE ADDHYD    CROSS SECTION                  3   OUTPUT HYDROGRAPH=  7
           INPUT HYDROGRAPHS=  5,6

                 PEAK TIMES        PEAK DISCHARGES        PEAK ELEVATIONS
                  12.45             214.332                 53.40
                  22.77              13.482                 50.54
                  23.91               8.970                 50.36

SUBROUTINE RUNOFF    CROSS SECTION  4  INPUT RUNOFF CURVE=  81.0    TIME OF CONCENTRATION=   .40
           AREA=    .12

                 PEAK TIMES        PEAK DISCHARGES        PEAK ELEVATIONS
                  12.13             160.405                (RUNOFF)
                  22.59              16.080                (RUNOFF)
                  23.61               2.970                (RUNOFF)

SUBROUTINE REACH     CROSS SECTION  5  INPUT COEFFICIENT=   .0000    INPUT ROUTINGS=     .00
           LENGTH=  1150.00    AVERAGE WATER VELOCITY=   4.470    AVERAGE ROUTING COEFF=   .7245    NUMBER OF ROUTINGS=    .52

                 PEAK TIMES        PEAK DISCHARGES        PEAK ELEVATIONS
                  12.52             211.962                 46.27
                  22.83              13.332                 44.36
                  23.97               8.962                 44.20

SUBROUTINE ADDHYD    CROSS SECTION                  5   OUTPUT HYDROGRAPH=  7
           INPUT HYDROGRAPHS=  6,5

                 PEAK TIMES        PEAK DISCHARGES        PEAK ELEVATIONS
                  12.25             301.137                 46.47
                  22.65              18.131                 44.40
                  23.95              11.919                 44.26

ENDCMP

ENDJOB CARD ENCOUNTERED.   END OF JOB.
```